X

Ice Drift, Ocean Circulation and Climate Change

Springer
London
Berlin
Heidelberg
New York
Barcelona
Hong Kong
Milan
Paris
Santa Clara
Singapore
Tokyo

Jens Bischof

Ice Drift, Ocean Circulation and Climate Change

Springer

Published in association with
Praxis Publishing
Chichester, UK

PRAXIS

Professor Jens Bischof
Department of Ocean, Earth and Atmospheric Sciences
Old Dominion University
Norfolk
Virginia
USA

SPRINGER–PRAXIS BOOKS IN ENVIRONMENTAL SCIENCES
SUBJECT *ADVISORY EDITOR*: John Mason B.Sc., Ph.D.

ISBN 1-85233-648-X Springer-Verlag Berlin Heidelberg New York

British Library Cataloguing-in-Publication Data
Bischof, Jens
Ice drift, ocean circulation and climate change. –
(Springer-Praxis books in environmental sciences)
1. Sea ice drift 2. Paleoceanography 3. Paleoclimatology
I. Title
551.3′42

ISBN 185233648X

Library of Congress Cataloging-in-Publication Data
Bischof, Jens, 1956–
Ice drift, ocean circulation, and climate change / Jens Bischof.
p. cm. – (Springer-Praxis books in environmental sciences)
Includes bibliographical references.
ISBN 1-85233-648-X (alk. paper)
1. Paleoceanography–Quarternary. 2. Sea ice drift. 3. Paleoclimatology–Quaternary.
I. Title. II. Series.

QE39.5.P25.B47 2000
551.46–dc21 00-044030

Printed by MPG Books Ltd, Bodmin, Cornwall, UK

Copy editing: Rachael Wilkie
Cover design: Jim Wilkie
Typesetting: BookEns Ltd, Royston, Herts., UK

Printed on acid-free paper supplied by Precision Publishing Papers Ltd, UK

Table of contents

Acknowledgements – Dank

All thanks are in random order, and here I am taking the freedom to switch between English and German as appropriate.

First I wish to thank the publisher, Praxis Publishing, for giving me the chance to write this book. The offer could not have come at a better time. It provided me with the opportunity to take a new look at seemingly 'old' data that now, in the light of the climate debate, suddenly look rather exciting. At least that's what I think. Also, some of the data and evidence presented in this book would probably never have seen the light of day, considering the page limitations of major scientific journals. Once again, thanks guys, for believing in me!

I want to express my gratitude to Peter Lemke of the Insitut fuer Meereskunde at the University of Kiel, Germany, for his chapter on the mathematical foundations between ice rafting and the surface circulation.

Clive Horwood: thanks for keeping me on my toes. It was necessary.

Thanks to the National Science Foundation and the Office of Naval Research, whose financial support has made my studies possible.

Dank aus demselben Grunde auch an die Deutsche Forschungsgmeinschaft und den Deutschen Akademischen Austauschdienst.

Der Universität Kiel sei gedankt fuer meine Ausbildung, ohne die die in diesem Buch dargestellten Ergebnisse nie das Licht der Welt gesehen hätten.

Jörn Thiede fuer seine Vision eines marinen Forschungsinstitutes und für die Chance, eine Doktorarbeit durchführen zu können.

Es geht natürlich weiter mit der Kiel Connection:

Robert Spielhagen, ein echter Freund, wenn auch als co-Autor ein wenig störrisch. Ich werde mich noch lange an unsere fachlichen und privaten Diskussionen erinnern. Ist nicht leicht, Dich zu überzeugen! Außerdem ist Robert bis auf diesen Tag der einzige, der gegen mich eine ausgeglichene Bilanz im Squash hat. Nicht schlecht, Robert!

Heidi Kassens und Elke Vogelsang; well, ich habe Euch einmal vergessen, was mir noch Jahre danach aufs Brot geschmiert wurde. Diesmal wird es mir nicht passieren. Deshalb also: Dank, Dank, immer wieder Dank! Im Ernst: vielen Dank fuer die mir zur Verfügung gestellten Proben, ohne die eine Bestimmung des zeitlichen Ablaufs des Zerfalls des Barents See Eisschildes nicht möglich gewesen wäre.

Gerhard Bohrmann, ein Freund und häufiger Squash Partner, fuer anregende Diskussionen und spätabendliche Squash matches

Horst Bohrmann: um Deine Frage zu beantworten: ja, ich trinke immer noch 'Export' ...

Jutta Giesler fuer gute Stimmung (Hopfenblütentee) auf Polarstern und dafür, mich erfolgreich von der Seekrankheit befreit zu haben

den Besatzungen der Deutschen Forschungsschiffe Polarstern, Meteor und Poseidon – dieses Buch enthält Ergebnisse von allen 3 Ausfahrten, an denen ich während meiner Doktorarbeit teilnahm, Meteor im Sommer 1986, Poseidon im Herbst 1987, und Polarstern im Sommer 1989. Beste Küche: Poseidon 1987, hilfsbereiteste Deckmannschaft: Polarstern 1989. Außerdem: es tut gut, wenn man nach getaner Arbeit mal einen 'zischen' kann (im Gegensatz zur 'trockenen' Atmosphäre auf amerikanischen Schiffen ...).

To the crew of USCGC 'Polar Star' for the successful trip in the summer of 1992.

Meine früheren Kollegen und ehemaligen co-Doktoranden Jens Matthiessen, Henning Bauch (der mich harsch daran erinnerte, das Fussball tatsächlich ein Kontaktsport ist), Ayrton Lackschewitz, Ingo Wollenburg, Uwe Pagels, Matthias Hüls, Simon Jung, Michaela Kubisch und Lovisa Birgisdottir.

Gudrun Thiel, die beste Sekretärin auf der ganzen Welt.

And now to my American collegues and institutions:

The University of Wisconsin at Madison for having me for my first postdoctoral stint. I enjoyed my one year stay there very much.

Old Dominion University in Norfolk for providing the lab space, offices and instrumental support necessary for me to conduct the research that fills a large portion of this book.

Dennis Darby for being a friend, mentor, and indirectly responsible for me finding my wife! Dennis, our collaboration is now entering its ninth year – may we never run out of good ideas ...

David Clark, for being a friend, a good influence, and for agreeing to review this book

I wish to thank the following individuals for supplying samples without which our source determinations would not have been possible. They are: John England, Donald Lemmen, Robert Gilbert, Weston Blake Jr., Peta Mudie, Iris Hardy, Jean-Serge Vincent, Svend Funder, Carl-Egede Bøggild, Anders Solheim, Robert Spielhagen, Scott Lamoureux, Arthur Dyke, Douglas Hodgson, and Bobbi Conard.

Robert Gilbert is thanked for allowing me to use his magnificent iceberg pictures . . .

. . . and for the same reason, I also thank Julian Dowdeswell: your pictures really do say more than thousand words!

Thanks to the anonymous person who suggested me as the author of this book; I think I know who you are . . .

And last but certainly not least, thanks to my family: my lovely wife Rosa, our children Christa and Karl, my parents Rosemarie and Günter, and my siblings Gerd and Gabi. It is good to hear words of encouragement when undertaking something as laborious as writing a book. Thanks und Danke Schön!

Foreword

Just before I left the University of Kiel in 1983 with my diploma, I overheard two students enthusiastically debating the possibility of reconstructing the past ocean surface currents by tracing the debris falling to the sea floor from drifting icebergs. I remember walking past those individuals, shaking my head in disbelief. Such an idea could be conceived only in the discussion-heavy intellectual climate of a university, perhaps in the early morning hours or after too much drinking. The idea seemed so outlandish to me that it seemed an impossible goal. Little did I know that only three years later I would find myself pursuing that objective. In the spring of 1986, I was looking for a doctoral thesis project, and tracing iceberg footprints suddenly sounded like a good idea. In the framework of the special research project 'Sedimentation in the Norwegian Greenland Sea' I selected ice rafting in the Norwegian Sea as my thesis topic and have, in essence, never left the field again. Since then, the idea of reconstructing ocean currents by following iceberg droppings has become more and more popular, and more investigations with focus on ice rafted debris are being conducted now than ever before.

The beauty of using ice rafted rock and mineral fragments in ocean sediments as a tracer of the past ice and surface water motion lies in the fact that it provides clear evidence of direction and movement, as long as the origin of the ice rafted particles can be accurately determined. In a simplistic way, icebergs and sea ice floes act as floats and drifters that occasionally mark their geographic positions by dropping parts of their bedload along the way. Studying the distribution of ice rafted debris, therefore, is a very powerful paleoceanographic tool and, combined with other sedimentary parameters, leads to an improved understanding of past ocean environments. And some of those environments, as will be demonstrated in this book, were profoundly different from those of the present. Even our ideas about the present-day climate and its origin (chronologic evolution) must be revised in some cases. If we succeed in determining the causes for oceanographic changes, which of course are linked to changes in the atmosphere also, the paleo-aspect of this type of analysis may attain predictive capabilities. In short, any computer climate model designed to predict future climate change such as greenhouse gas-induced global warming will have to reproduce the reconstructed past changes of the ice drift as well in order to be considered reliable.

From attending scientific conferences and workshops – some of the latter intended to chart the course of future research – I received the impression that only a few scientists fully comprehended the power of tracing iceberg driftpaths for climate reconstructions and for the development of predictive models. Many scientists active in the field of climate research are not even aware that such studies have been performed and that the results are published and widely available. Geochemical studies, using highly sensitive and expensive instruments, outnumber by far the comparatively few studies obtained with traditional geological methods such as looking through a microscope. Therefore, this book was written in the hope that it may convince a larger number of scientists about the usefulness of 'following iceberg's footprints' as Eystein Jansen once called it (Jansen, 1992), as not just a complementary, but an essential and very powerful tool for paleoceanography.

This book presents a summary of what we know and do not know about ice rafting and its connection to the ocean currents and global climate. It is hoped that it will stimulate further research and perhaps open up a few new avenues not thought about so far. The reader should bear in mind that not all ice-covered seas are equally well studied with respect to ice rafting, and that the scope and focus of the investigations are not the same. This makes this book somewhat sketchy, at least from a geographical point of view, and there is a bias towards the area of expertise of the author, the high-latitude northern hemisphere.

Jens Bischof, Norfolk, Virginia

About this book

From discussions and conversations with fellow scientists in the field of paleoclimate research, I received the impression that the concept of using ice rafted debris as an instrument to reconstruct the past ocean surface circulation is not very well understood. In fact, many scientists are so specialised in their own disciplines that they are completely unaware of the connection between ice rafting and climate change. Therefore, the time has come for a definitive book on this subject. This book was written for anyone involved and/or interested in climate research. It is suitable for university students from graduate level upwards, as well as for scholars of global climate. Climate research is a hot subject because of the prospect and threat of anthropogenically-induced global climate change. This book focuses on interactions between the ice, the ocean and the atmosphere, and describes how these three components of the climate system influence each other.

Ice rafting, the drift of floating ice in the ocean from one place to another, is not just a passive recorder of the past surface ocean circulation, it also actively influences and changes the surface ocean circulation in ways that are described in this book. Ice rafting has left an almost uninterrupted archive of past surface circulation changes in deep-sea glaciomarine sediments. Wherever ice forms in contact with the land, whether as icebergs calving from a glacier that overrode and incorporated bedrock fragments into the ice, or as sea ice that forms when sea water freezes in contact with unconsolidated sediments in coastal environments, terrigenous debris from the ice's place of origin is embedded in the ice. Along the following drift in the ocean, the incorporated particles drop to the sea floor when the ice melts or when the icebergs break up or turn over. In any case, all incorporated debris is lost between the origin of the piece of drifting ice and the position at the end of its lifespan. Over the course of decades, centuries and thousands of years, a continuous archive of iceberg and sea ice drift has formed in the deep-sea sediments. The petrographic composition of the ice rafted debris (IRD) in these sediments reveals the place of the ice's origin and allows a reconstruction of the surface currents of the past. One method is to compare the IRD petrographies to known bedrock formations in potential source areas. Another, better, possibility is to match the IRD composition to the composition of debris from previously-determined source areas using statistical methods. Both approaches are presented in this book.

Since the motion of icebergs and sea ice is controlled by the atmospheric and oceanic circulation, the dispersal paths of IRD from known sources permit the reconstruction of the past ocean surface currents and winds.

This book is organised into essentially three sections. The first section introduces the concept of ice rafting. Most scientists today are highly specialised in their own disciplines and don't fully understand the concept of ice rafting. Therefore, the different types of ice, the processes that lead to the incorporation of lithogenic debris into the ice, and the process of rafting – drifting from one place to another across the ocean – will be defined. The second section is devoted to ice rafting in the high latitude northern hemisphere oceans. The composition of ice rafted debris in Late Quaternary sediments of the Norwegian Greenland Sea and the Arctic Ocean will be described first. Then, the IRD will be traced back to its origin using a variety of methods. This particular section is the flesh and blood of this book. It contains a high level of detail, which may make some passages challenging to read. But this detailed discussion, and a careful introduction to the thought process that leads to the source identifications, is crucial to convince the reader that the source determinations are indeed correct and trustworthy. Throughout this section, pieces of evidence will be strung together to form a model of the late Pleistocene and Holocene surface circulation and late Pleistocene ice sheet dynamics. Also, the physical driving forces of this circulation are discussed.

The next chapter is devoted to the mathematical foundations of the connection between ice drift, surface ocean and atmospheric circulation. It was written by my guest author, Dr Peter Lemke from the Institute of Marine Sciences at the University of Kiel, Germany.

Finally, all information on IRD will be fused into a model of the ocean surface and atmospheric circulation during the last 1.5 Ma (Arctic Ocean), and the last 400 ka (Norwegian Sea). It will be demonstrated that the climatic variability during the Late Quaternary was enormous.

Some highlights of this book are that

- the northernmost part of the North American Arctic was glaciated at least 13 times during the last 1.5 Ma;
- an ice sheet covered the Barents Sea not only during the Late Weichselian, but also at least four times before;
- the surface currents in the Norwegian Sea generally switched from northwards during interglacials to southwards during glacials;
- the present climate in the Norwegian Sea is unstable and punctuated by brief returns to southward, cold, icy currents instead of northward, temperate currents;
- the Arctic Ocean circulation oscillates between two dominating and atmospheric driven modes; and
- the volumes of icebergs that calved from the Pleistocene ice sheets in part depended on temperature changes, but in part on internal ice sheet dynamics.

This book is not a compilation of published articles but presents new data, and a new look at older data that otherwise probably would have never seen the light of day. It also contains personal views and opinions, which are clearly marked as such wherever they appear. There is a bias towards the northern hemisphere, not because the circum-Antarctic sediments are less well studied, but because the focus of Antarctic studies is different from the focus of this book. I am unaware of IRD provenance studies at greater distance from the Antarctic continent.

It is my hope that this book will convince many, who at the moment are unaware of the concept of tracing icebergs' footprints in order to decipher the past ocean circulation, about the usefulness of IRD provenance studies as a very powerful paleoceanographical and paleoclimatological tool.

1

Introduction

Climate change has become a topic of great public interest. Hardly a week goes by without newspaper articles proclaiming global warming, the greenhouse effect, melting polar ice caps, and retreating glaciers. No self-respecting weatherman can resist the temptation to see a connection between slightly abnormal weather patterns and EL Niño,[1] the eternal culprit. While it is clear that the burning of fossil fuels (petroleum, coal, wood) and the ensuing rise of the carbon dioxide level in the atmosphere must trigger a reaction of the global climate system, it is completely unknown what kind of reaction will occur. As a colleague of mine (Richard Alley) once put it, 'we are performing the greatest experiment with the global climate right now'. And indeed, there are signs from some natural systems, the circum-Arctic regions in particular, that global warming is under way: observations of the pack-ice thickness of the Arctic Ocean from submarines with upward-looking sonar show a thinning trend since the 1970s, the margin of permafrost is moving north, and the vegetation in the high northern parts of the world is changing towards more temperate forms (the so-called 'greening' of the Arctic). But it is by no means clear whether these signs indicate real, worrying signs of man-made permanent and potentially disastrous climate change, or just regular, naturally-occurring variations of the Earth's climate system. Clearly, the answers to these questions can only be found in the paleo-record of marine sediments, in which climate changes over the course of long periods of time are preserved as an uninterrupted archive.

If proven a reality, the most troubling aspect of global warming is that it would cause melting of the polar ice caps, which in turn would cause the global sea level to rise, and flood some of the most densely populated regions on Earth. One does not need to be a prophet to imagine the consequences; forced emigration of unprecedented scale into higher elevated areas would strain the economies and societies of the involuntary host nations, causing political turmoil and, knowing how humans traditionally react to such changes, most likely war.

Other effects related to global warming could be changes in the rainfall patterns, which could lead to widespread droughts and threaten agricultural production.

Because of these potentially disastrous consequences of global warming, intense study of the global climate system is important. We need a solid understanding of the natural principles that govern the climate and control its change. There are at least three ways to do this:

(1) straightforward climate modelling (global climate models, GCMs[2]) with the help of today's powerful computers, fed with physical parameters as far as we know them;
(2) synoptic observations of the present climate with expeditions such as the first major scientific crossing of the Arctic Ocean in 1994 (Wheeler, 1997), year-round climate observatories (for example the SHEBA project[3]), and from earth orbiting satellites;[4] or
(3) analysis of marine sediment cores with focus on past climate changes.

Each of these approaches has advantages and problems. The advantage of method (1) is that, at least theoretically, it would make it possible to predict future climate changes, and thus buy mankind (hopefully) sufficient time to prepare for the consequences. But realistic parameterisation of the physical forces that drive the climate, in particular of the cloud cover, remains problematic and adds insecurity to the model outcome. The virtue of method (2) is that it produces much needed ground-truth data of the present-day climate. The SHEBA data, for example, eliminated the need to make educated guesses about parameters such as air, ice and water temperature, albedo, ice cover and thickness, solar radiation, snow cover, etc. in the high Arctic. These values are now precisely known for a whole year. But the problem with these 'climate snapshots' is that we do not know how representative these data are over longer time spans. Are the SHEBA data, just to name an example, 'unusual' (= beyond the range of natural variability) or 'normal' (= within the range of natural variability)? Even if the temporal scope of climate observations is extended to several decades, one still does not know how the observed changes compare to those that have occurred over the course of hundreds or thousands of years. Later in this book an example will be presented that illustrates how misleading our ideas of climate change can be if the observational time period is not long enough.

The complete answer can only be found with method (3). Deep-sea sediments are for the most part well protected from erosion and contain uninterrupted archives of climate changes on various time-scales. By analysing the sediment composition for a wide array of parameters (ice rafted debris, biogenic vs. terrigenous material, microfauna and -flora, sediment geochemistry, isotopic composition, etc.), one can obtain information on past water temperatures and their annual fluctuation, the quantity and species distribution of organisms that populated the ancient waters, and on the extent of ice cover and icebergs. But there are problems: the parameter determinations are, of course, not as accurate as those obtained with method (2), the sediment record can be corrupted by secondary (post-depositional) processes such as carbonate and silica dissolution (which can make a sediment look as if it was deposited in a biological desert, when in fact the seas were bustling with life), bioturbation (the burrowing action of organisms that live on the sea floor), and deep-water and bottom currents. Furthermore, the sedimentation rates set limits on the time resolution that can be obtained from marine sediment cores. In the high latitude oceans, sedimentation rates are in the order of only a few centimetres per thousand years (in the Arctic Ocean, they are even lower, only a few millimetres), so that every sample integrates over hundreds if not thousands of years.

As we have seen, no one single method can produce all the required information to fully understand the global climate system. But combined, these methods will lead to a sounder understanding of the global climate and enable us to predict its future development better than now. Specifically climate-oriented research is now being conducted more intensely than ever before, and we are continuously learning more about how the Earth's climate works.

What makes the study of the present climate so complicated are the multiple feedbacks between its main elements: the ocean, the atmosphere and the hydrologic cycle. A good example for this is the North Atlantic Region. There, the North Atlantic Drift[5] pumps large volumes of water and heat northward in the eastern North Atlantic. These warm waters release moisture and heat to the atmosphere and create a temperate and relatively wet climate in northwest Europe with ice-free conditions even north of Norway, while the same latitudes further west are under much colder conditions. Therefore, the ocean currents control the atmospheric temperature and moisture content. But conversely, the atmosphere is controlling the ocean circulation with mainly westerly winds over the North Atlantic.

Another element of the climate system is freshwater, which controls the density structure of the ocean and its freezing point. Freshwater exists as water vapour in the atmosphere, liquid water on the continents, as snow and ice on the polar ice caps (Antarctica and Greenland), and as pack ice and icebergs in the Arctic Ocean and around the Antarctic continent. All elements of the climate – the ocean, the atmosphere and the hydrosphere – both influence each other and have a monitoring function. Drifting ice responds to the motion of the upper ocean and the prevailing winds, very much like a Lagrangian drifter. Therefore, the net ice-drift direction is a vector product of both the atmospheric and oceanic circulation. Since ice can carry rock and mineral fragments of up to boulder size, and because these particles drop to the sea floor as the ice drifts and melts, ice leaves a trace of its drift path in marine sediments. The trace consists of lithologies that can indicate the location of the ice's origin. The process of ice rafting has thus created a nearly uninterrupted archive of the changing surface ocean and atmospheric circulation in Late Cenozoic high-latitude marine sediments.

This book will demonstrate how strongly the ice drift, and therefore the ocean and atmospheric circulation, changed during the last several 100,000 years. It will document the fact that some major climate changes of the past, for example the change from the last ice age to the recent Holocene, were preceded and accompanied by major changes in the directions of the ice drift, which themselves mark major perturbations of the ocean-ice-atmosphere system. But the ice rafted record reveals more than the history of circulation changes: it can also be used to infer the existence of past ice shields and the history of their growth and decay. The Barents Ice Sheet, for instance, existed in nearly the same configuration five times during the last 450,000 years, and the Laurentide Ice Sheet existed at least six times during the last one million years. Such conclusions are possible, because the great lithologic differences between the landmasses surrounding the high northern seas facilitate clear discrimination of major ice rafted debris (IRD) sources. From the first beginnings in the Norwegian Sea, the methods of tracing IRD back to its sources have become more sophisticated. It is now possible to trace individual sand grains, and not just entire grain assemblages, back to their sources and to determine the degree to which every sample represents a mixture from different sources.

This book is organised to give readers unfamiliar with the subject of ice rafting an overview of the most important physical processes that are involved in ice rafting, proceeding to actual examples of interpreting the ice rafted record in terms of paleocirculation and paleoclimate, and concluding with a summary of what we have learned thus far from the ice rafted record. Those who are familiar with and professionally involved in the study of ice rafting will probably find the later chapters more important.

First, the different types of ice and their capacity to carry rock and mineral grains will be introduced. In order to correctly trace ice rafted sand grain assemblages back to their sources, it is important to distinguish between sea ice and iceberg transport. The reader will realise that this distinction is not always possible. Then, a short theoretical treatment of the particle transport and release mechanisms by drifting ice is given. The main portion of this book is devoted to the presentation of the ice rafted records of the Norwegian and Greenland seas, the Fram Strait and the Arctic Ocean. Selected, individual chapters will be synthesised into a glacial history of the high northern hemisphere oceans as derived from the ice rafted record. The north polar regions are in this respect more interesting than the circum-Antarctic waters owing to the strong meridional exchange of waters and ice between the Arctic and the lower latitude oceans. By contrast, the circum-Antarctic waters move around the Antarctic continent in coast-parallel belts and thus do not extend far into the lower latitudes. Next comes a mathematical concept of the connection between the reconstructed ice drift directions, the surface ocean and the atmospheric circulation. Finally, major results are presented as a bullet list. It is hoped that this book will enhance research on climate change.

2

Historic perspective

From the first time that sediment samples were dredged from the sea floor of the high latitude oceans, almost invariably they were found to contain large pebbles and cobbles, whose presence could not be explained any other way than by ice transport. Scoresby was probably one of the first who described enormous quantities of rocks to be present on the surface of drifting icebergs (Scoresby, 1823). Later, Washburn (1947) recognised the importance of sea-ice rafting of large stones in the southwestern Canadian Arctic islands and concluded that ice rafting is an important sediment transport process in the Arctic. In his book, beautiful descriptions of various forms of particle (sand, clay, stones, pebbles) inclusions in and on sea ice can be found, and he proposed a variety of processes by which ice can incorporate, and then transport, heavy loads of lithogenic debris throughout the ocean. But it was not until the early 1970s that someone actually tried to link the ice rafted rock fragments to their most likely continental sources. Kudrass (1973) proposed that carbonate erratics in deep-sea sediments off the coast of Morocco were ice rafted from the coastal mountains facing the Atlantic. Ruddiman (1977) reconstructed the Pleistocene surface circulation of the North Atlantic on the basis of thickness changes of several distinct sand layers. Clark et al. (1980) suggested that the omnipresent carbonate clasts in Arctic Ocean sediments required a source in the Canadian Arctic or North Greenland. But the first attempt to reconstruct long-distance particle transport by ice rafting in the ocean by direct comparison of the erratic pebbles and gravels with rocks from their potential origin was performed by Wohlfeil (1983). He found Rapakiwi Granites and Rhomb Porphyries, very conspicuous and rare rocks known to occur in only two small localities in Scandinavia, among gravels dredged from the Iceland-Faeroer Ridge, and concluded that at one time ice must have drifted from the northeastern North Sea into the Norwegian Sea and then southward across the Iceland-Faeroer Ridge into the North Atlantic. Heinrich (1988) discovered that ice rafting around the Dreizack Mountains in the North Atlantic produced several distinct coarse layers, and suggested that Pleistocene ice rafting was cyclical and basinwide, an idea that later became known as the 'Heinrich Layers'.

After the publication of paleo-temperature records of Greenland ice cores (Dansgaard et al., 1993; Clark et al., 1999), a deluge of papers appeared exploring possible linkages between temperature changes over Greenland and the occurrence and abundance of ice rafted debris in the high latitude northern oceans. Bond and Lotti (1995) showed that the coarse ice rafted debris layers followed cold snaps in the Greenland ice record.

Glaciomarine sediments, the ultimate product of ice rafting, have been studied in detail for some time. They are characterised by variable quantities of coarse – in some cases up to boulder-sized – stones and sand in an otherwise fine-grained clay to silt matrix. The ice rafted stones, sometimes referred to as dropstones as a result of their depositional history, can be in random, chaotic orientation or in distinct layers. Biogenic components in glaciomarine sediments are often cold water species assemblages. Deep-sea sediments that were deposited in environments without ice lack the coarse lithogenic components and are composed of fine-grained, slowly settling clay and silt from distant sources and authigenic biogenic components.

Bruce Molnia presented a comprehensive volume on glaciomarine sedimentation (Molnia, 1983). The book contains examples of sedimentation in fjord environments, relatively close to the glacier (Powell, 1983), and from distant sites in the ocean (Minicucci and Clark, 1983). There are also examples of glaciomarine sedimentation from around Antarctica (Anderson et al., 1983), and examples of glaciations with glaciomarine sediments from earlier, pre-Cenozoic periods of the Earth's history (Anderson, 1983). Highlights of this book are that sedimentation rates are highest near the ice margin and then rapidly decrease with distance. Another very important factor that controls ice rafted sedimentation rates is the surface water temperature in which the icebergs drift. Two opposite extremes exist: the very cold, permanently ice-covered Arctic Ocean in which sedimentation rates are very low (1 mm in 1000 years and even less; Minicucci and Clark, 1983), vs. temperate, mid- to high-latitude fjord settings where glaciers terminate into warmer waters and release enormous amounts of glacial marine sediment. A very important implication of these findings is that icebergs in cold, ice-covered waters such as the surface waters of the Arctic Ocean can survive for long periods of time without experiencing significant melting and thus travel long distances, while icebergs released into temperate waters have a comparatively limited life span and do not drift very far.

Powell (1983) provides an excellent account of sedimentation in a temperate setting. Syvitsky (1989), Syvitsky and Hein (1991), and Andrews et al. (1991) describe the style of sedimentation in ice proximal sub-Arctic environments from the eastern Canadian Arctic Archipelago. The processes that lead to debris entrainment into glacier ice are described by Drewry (1986). This book contains excellent pictures of enormous sediment concentrations in the basal portion of Antarctic glaciers. Newer information on glaciomarine sedimentation can be found in Dowdeswell and Scourse (1990). For further reading on the subject of glaciomarine sedimentation, the reader should try *Geological History of the Polar Oceans: Arctic versus Antarctic*, edited by Ulrich Bleil and Jörn Thiede (1990). For more on the physical oceanography of the polar oceans, I recommend *The Polar Oceans and Their Role in Shaping the Global Environment* by Johannessen, Muench, and Overland (1994). More on the physical foundations of glacier movement can be found in Paterson's book *The Physics of Glaciers* (1994).

The work of this author has concentrated on the directional aspect of ice rafting and, thus far, has produced these results among others:

- the surface currents in the Norwegian Sea moved in the opposite direction than today from circa 17,000 to 15,000 years before the present and several times during pre-Weichselian glaciations;
- the East Greenland Current has shifted from a position near the coast of East Greenland to the centre of the Norwegian Greenland Sea frequently during the Pleistocene, most notably during marine isotope stage 6, from 180,000 to 130,000 years b.p.;
- the Barents Ice Sheet existed and disintegrated at least five times during the last 450,000 years;
- the Arctic Ocean ice drift has been very variable through time – its main element is a Transpolar Drift (TPD[6]) that carries ice across the North Pole towards Fram Strait into the Greenland Sea. At times the TPD drew more from northern North America, at times more from Eurasia, and at times from both sides of the Arctic, resulting in maximum volumes of ice export to lower latitudes. The observational basis and the arguments leading to these conclusions will be presented in the appropriate chapters of this book.

3

The concept of ice rafting

> 'A survey of the means by which rock debris may become ice rafted, observations of debris in process of being carried by ice, and observations of the results indicate that ice rafting is a process of considerable quantitative importance...'
>
> Albert Lincoln 'Linc' Washburn, 1947

The conceptual idea behind ice rafting is remarkably simple: floating ice that contains lithic particles (stones, pebbles, fine sand grains) loses these particles episodically during its drift in the ocean and thus marks its drift track with rock and mineral fragments indicative of its source.

Wherever ice is in contact with terrigenous debris, whether it be clay, sand, gravel, boulder clay etc., the potential for entrainment and enclosure of the lithic debris exists. Alpine glaciers tend to carry a debris load at the bottom (ground moraine), in front, at the sides, and in englacial positions, whereas polar ice caps such as the large Antarctic and Greenland ice sheets and their Pleistocene precursors have debris at the underside (Drewry, 1986). Glaciers essentially incorporate lithic debris by overriding bedrock or sub-glacial sediments. The pressure and forward motion of the ice cause the underlying rock to break up into pieces, which subsequently become incorporated into the bottom of the ice stream. Then the erosive action of the ice is amplified by the freshly crushed, angular rock fragments at the glacier sole, which greatly increases the friction between the glacier and the underlying bedrock. Once the glaciers reach sea level, icebergs calve off and start to drift in the ocean, beginning the process of ice rafting.

Sea-ice that forms from freezing saltwater is also capable of particle entrainment in a variety of ways, which is discussed in detail in Section 4.2 on particle incorporation. As glaciers flow to the sea level, they either form a floating ice shelf (such as in the large Filchner and Ross ice shelves in Antarctica), or icebergs calve off the front directly (as, for example, in the large fjord systems of Scoresby Sund, East Greenland, Jakobshavn Glacier, West Greenland and the fjords of West Spitsbergen), depending on the physical setting of the glacier terminus. Either way, iceberg calving is an inevitable consequence of glacier ice reaching sea level. As the icebergs start drifting in the ocean, they lose the attached debris particles gradually due to the pull of gravity, or episodically when the icebergs melt and roll over. In the case of roll over, debris that was attached to the underside of an iceberg can appear on the top and be washed off the surface during the summer melting period, as observed by Gilbert (1983, 1990). Once released, the heavier particles (sand grains and coarser material) drop nearly vertically to the sea floor and thus mark the iceberg passage with rock fragments from the iceberg's origin. All a geologist has to do to reconstruct the past ice drift directions is to find a method by which the particles can be connected with a high level of confidence to a specific source region. In this book, various approaches will be presented and discussed.

4

Ice rafting and climate change

What are the multiple connections and feedback loops between drifting ice, ocean circulation and the climate system? The process of ice rafting is intimately connected to temperature changes on global and regional scales. Cooling in the high latitude (north of 60°N and south of 60°S) regions leads to an expansion and thickening of the sea ice cover in the oceans, and this pushes the polar fronts toward the Equator. The process operates on different time scales: cooling happens seasonally (winter and summer), but also on much longer time scales of thousands and tens of thousands of years – the ice age cycles. Cooling would also increase the amount of snowfall. These combined processes increase the 'whiteness' of the polar regions (the aereal extent of snow- and ice-covered regions) and thus increase the albedo.[7] And the albedo increase, in turn, further amplifies the cooling trend, creating a positive feedback between cooling, leading to more ice and snow, higher albedo, and more cooling.

A similar effect can be achieved by physically moving excessive amounts of ice from the polar regions to lower latitudes. If, for example, the air pressure distribution over the Arctic Ocean were such that winds blew from the Bering Strait across the Pole towards Fram Strait, then massive amounts of pack ice would be moved into the Norwegian Greenland Sea and push the polar front southward. In the winter, this process would continuously produce additional sea ice in the open leads created by the offshore winds in the Bering Strait region, setting in motion a veritable 'ice machine'. The regional extent of ice and snow cover in the Norwegian Greenland Sea would increase, cooling the region, and the high albedo would reflect more of the incoming solar radiation, further amplifying the cooling. Depending on the strength and duration, this process could lead to an episode of relatively cold climate over the North Atlantic region, perhaps lasting from a few years up to decades. But if it were sufficiently strong and durable, it could set the stage for the global climate to return to full glacial conditions.

The difference between these two cooling processes is that the first requires an external (orbitally-forced[8]) motor of cooling while the second (the different ice drift) does not. In the second process, a mere change of the ice drift direction in the Arctic region could set cooling on its way, possibly even on a global scale. Such changes between a relatively warmer world as today, and glacial climates during which much of the earth was submerged under thick ice sheets, have occurred repeatedly during the Late Pliocene to Pleistocene.

In the high latitude regions around both poles, the traces of glaciation – glaciomarine sediments with ice rafted coarse-grained debris – can be found only in marine sediments younger than mid to late Pliocene (northern hemisphere), or younger than mid Tertiary (southern hemisphere). In an uplifted glaciomarine sequence from the Gulf of Alaska, the Yakataga Formation, the onset of ice rafting dates back to the Miocene (Eyles et al., 1991). Even though the details are somewhat sketchy, it is now believed that the south polar glaciation began about 30–35 million years ago (Andersen, 1990), while the northern polar regions lagged behind that. There is uncertainty about the age of the onset of northern hemisphere glaciation. According to Wolff and Thiede (1991), it began at 3.6 Ma, but other authors (references in the same publication) give ages ranging from 4.2 to 2.5 Ma. Now, consensus seems to have been reached for a northern hemisphere initiation date of approximately 2.6 Ma (Clark, pers. com.). It should be kept in mind that these dates are based on the first occurrence of coarse grained, ice rafted rock and mineral fragments in deep sea sediments after long periods of steady fine grained and mostly biogenic sedimentation (Krissek, 1989). Lithic particles over 63μm in size indicate ice transport because surface ocean currents are not strong enough to carry these particles far away from their continental origin. The only question mark remains about the type of ice that carried the grains. Coarse ice rafted debris is often seen as direct evidence of iceberg rafting, a tempting interpretation because it would allow one to infer the existence of continental glaciers and ice sheets. However, this idea overlooks the fact that other types of ice, such as sea ice, ice that forms from freezing sea water, river ice, or landfast ice that forms in estuarine and fjord environments, also can transport coarse particles over significant distances (Spjeldnaes, 1981). Without any additional evidence for glacial ice rafting in the form of glacially-striated rock surfaces, physically crushed quartz grains, or an associated change in the composition of the marine microfauna and flora from warm or temperate to cold water forms, the conclusion that terrigenous >63m particles indicate iceberg transport remains ambiguous. However, recent results from a high resolution sediment core from the western Arctic Ocean suggest that three criteria, the synchronous drop in (1) the >250μm IRD weight percentages, (2) the lithologic diversity of the IRD, and (3) the maximum size of the IRD particles together define the end of the glacial era, and thus the end of iceberg rafting (Fig. 4.1; Bischof et al., 1999). Therefore, IRD grain frequencies of less than 10 grains >250μm per gram sediment, a maximum grain size limit of 2 mm, and a petrographic diversity of less than 30 different IRD grain types are herewith suggested as proximal sedimentological criteria that define sea ice rafting.

The global climate was warm, or at least temperate, for most of the Mesozoic period and began to cool during the Tertiary. From the middle Tertiary on, the Earth entered a glacial mode with the presence of continental glaciers and ice sheets and drifting sea ice, first around the southern polar regions and later in the north. The severity of the glaciations underwent several stages of intensification and, after approximately 1.6 Ma, the Earth has experienced rather regular cycles of glacial and

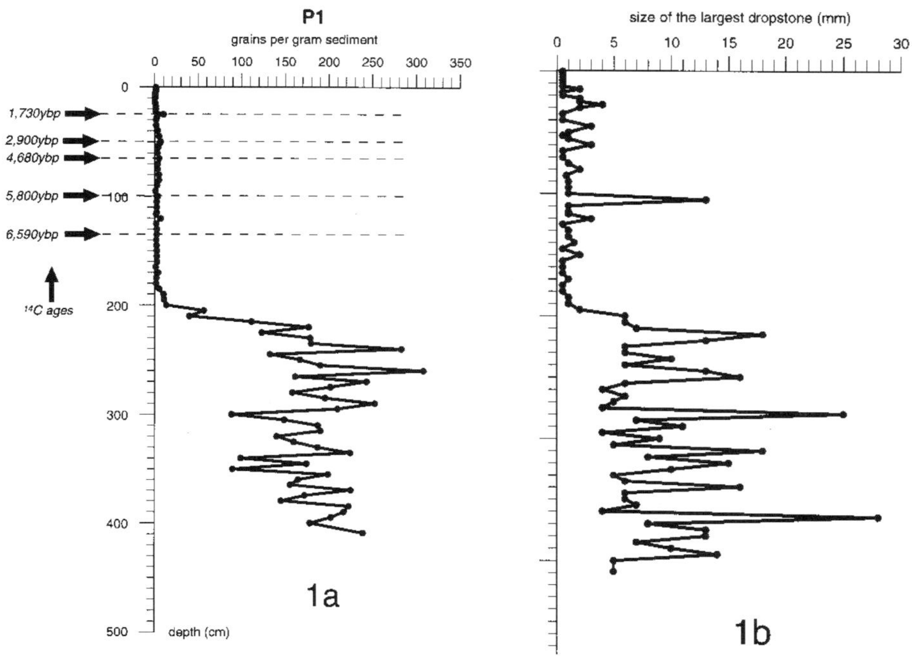

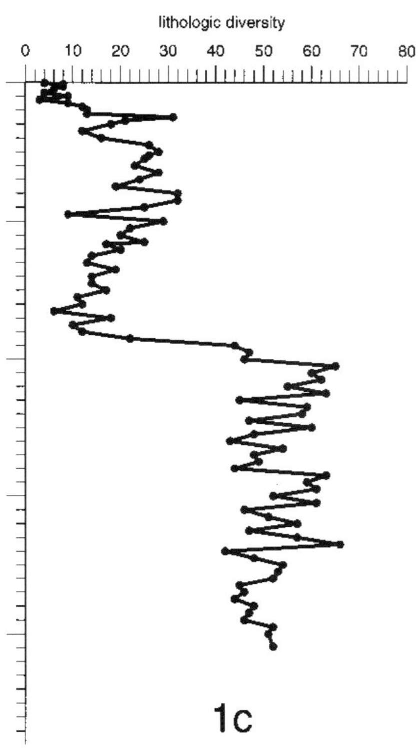

Fig. 4.1. Textural IRD parameters defining iceberg rafting are large grain frequencies of ice rafted lithic grains >250μm in size (1a), large maximum size of the dropstones (1b), and great lithologic diversity (1c). All these parameters drop synchronously at the boundary between the Late Pleistocene and the Holocene in circa 200 cm depth, which characterises the transition from glacial iceberg rafting to interglacial sea ice rafting. The example is piston core P1 from the western Arctic Ocean. Sea ice rafting (above 200 cm) is characterised by less than 10 grains >250μm per gram sediment, a maximum IRD particle size of generally 2 mm, occasionally more, and a petrographic diversity of less than half the Pleistocene values. It should be noted that, with only one exception, the maximum values for all three parameters for Holocene sea ice rafting are below the minimum values during the Pleistocene portion of the core.

interglacial periods with continental ice sheets and sea ice in both hemispheres. In this mode, relatively long-lasting (70–100kyrs) glaciations during which much of the northern hemisphere's landmasses were buried under laterally extensive and thick ice sheets alternated with shorter interglacials similar to the climate of the present. But if we consider that even now large parts of the circum-Arctic regions (Greenland, Svalbard, Franz Josef Land, Zevernaya Zemlya, the northeastern Canadian Arctic islands) and all of Antarctica are covered by several kilometre thick ice sheets, it is really appropriate to call the present state of the Earth's climate a glacial mode. In this mode, times with reduced ice cover alternate with times of widespread glaciation like the last glacial maximum, 18 kyrs before the present.

4.1 SEA-ICE PROCESSES

Sea ice forms when seawater cools owing to heat loss to the atmosphere. In the northern polar regions, this process begins around September or October when surface air temperatures drop below zero. The ice thickness continually grows and reaches 1 to 1.5 m by the end of April in the marginal parts of the Arctic Ocean (first year ice; Reimnitz and Barnes, 1975; Pfirman et al., 1990). Ice thicknesses are not that large in the circum-Antarctic waters where the average ice thickness is only 0.5 m (Geiger et al., 1998). The major ice 'factories' around the Arctic Ocean are the vast Siberian shelves, the Chukchi Sea, the Barents Sea and the Canadian Arctic Archipelago. In these marginal regions, most if not all ice melts in the summer before a new freezing cycle begins. At the present, approximately one half of the Arctic sea ice disappears during the summer but refreezes during the following winter.

Because of the climatic sensitivity of the high Arctic region, the physical condition of the sea ice cover has become a subject of intense scientific scrutiny. One reason for this is that, if global warming really is happening, it should be seen first as a thinning and shrinking of the Arctic sea ice cover. The other main reason is that a shrinking sea ice cover would probably amplify the warming through the decreasing albedo. And recently it was found that the aerial extent of Arctic sea ice is indeed shrinking (Cavalieri et al., 1997; Johannessen et al., 1995).

The eventual history of an Arctic sea ice floe is not just temperature dependent but also depends on its drift direction, velocity and sediment or particle content. During a 'normal' annual cycle, ice freezes in the winter through addition of ice to the underside of the newly formed or older sea ice floe. The thickness gradually increases until melting begins in the spring. Because of the increasing amounts of solar radiation, the ice melts at the surface and melt ponds begin to form. These melt ponds of light greenish to light blue colour have a lower albedo than clean ice or snow and thus trigger more melting. Then melting can be further amplified by sediment and/or biogenic material (diatoms and brown algae) in the melt ponds. Beautiful images that show what this annual cycle looks like in real life can be viewed on the SHEBA web site http://sheba.apl.washington.edu/images).

But if the sea ice drifts north, away from the marginal areas of summer melt, it can survive several melting seasons in the permanently frozen part of the Arctic

Ocean. In this case, some melting would occur at the surface and lead to a reduction of the ice thickness but not to the destruction of the floe, owing to the lesser amounts of solar radiation received at these locations. In the following winter, the sea ice floe would grow again downward due to addition of ice to its underside (Pfirman et al., 1990). Depending on the residence time of the floe in the very cold parts of the Arctic, this process can repeat itself and lead to a thickening of the ice and to an aging of the floe (multi-year ice). On average, multi-year ice floes in the Arctic ice cover have a thickness of 3 m, but this can be variable from year to year. Wadhams (1983) noticed a decline of the ice thickness during a 20 year-long period during which ice thicknesses were measured by upward-looking echo sounders from submarines. Some of these changes could be due to rising global temperatures.

But what complicates the straightforward interpretation of thinning sea ice as an indication of global warming are difficulties in separating the effects of anthropogenic causes such as increasing greenhouse gas levels in the atmosphere from naturally-occurring climate oscillations. For example, we (Ingo Wollenburg, Stefanie Pfirman, Peter Goldschmidt and I) drilled boreholes through sea ice in the eastern Arctic Ocean and the Barents Sea during the Polarstern expedition ARK VI/1–2 in 1989 and found multi-year ice thicknesses of mostly over 3 m, with maximum thicknesses of 4.3 m (Krause et al., 1991). Ice thicknesses of the SHEBA experiment were much thinner because they were measured mostly in thinner first-year ice. However, that experiment finished with a net loss of ice (Perovich et al., 1999a, b). It is therefore interesting to register that, 100 years earlier, Fridtjof Nansen measured ice thicknesses at the end of the Arctic winter generally in the range of 2.3–2.8 m in essentially the same area that we visited with the resarch vessel Polarstern 95 years later (Nansen, 1898). Because a strong anthropogenic signal can be ruled out for the time period of Nansen's expedition, his measurements probably indicate naturally-occurring ice thickness variations. Naturally-occurring, and quite drastic, changes of the extent of the Arctic ice cover and the surface water temperature are also indicated by changes in the palynomorph assemblages and their stable isotope signatures. This as yet unpublished work was carried out by my collegues Anne deVernal and Claude Hillaire-Marcel as part of a collaborative research project under the umbrella of the new SBI initiative (Shelf to Basin Interactions of Arctic System Science) sponsored by the National Science Foundation (Darby et al., 2000). Compared to the large area covered by ice in the Arctic seas (approximately 13 million square kilometres), the circum-Antarctic waters are covered by a narrow belt of much thinner sea ice (Geiger et al., 1998).

The Arctic sea ice cover plays an important but poorly understood role in the heat budget of the northernmost parts of Earth. Its aerial extent and thickness are, and most likely have been, highly variable as a result of factors such as air temperature, drift direction and velocity, and terrigenous and biogenic particle content.

4.2 PARTICLE INCORPORATION

The possibility of particle entrainment into or onto sea ice exists where it is in contact with land and on shallow shelves. If sea ice forms in fjords with steep walls, for example in western Norway, around the coast of Greenland, or in northern Ellesmere Island, talus debris can fall from the mountain slopes onto the ice. Clearly, this is a very effective process that can raft enormous quantities of sediment some distance. A collegue of mine (Uwe Pagels) observed a gravel and boulder accumulation on a sea ice floe more than 50 km off the east coast of Greenland. The gravel pile looked as if it had been dropped by a truck and clearly represents such talus debris.

In shallower environments with less relief energy, fine-grained particles can be blown from exposed land surfaces onto the ice surface by winds. When the spring breakup of rivers draining towards the Arctic Ocean precedes the summer melting of the Arctic ice pack, river sediment can be spilled on the ice pack. The annual freeze–thaw cycles can cause unconsolidated sediments (for example Late Wisconsinan glaciomarine sediments and/or terrestrial Wisconsinan glacial tills) from coastal cliffs to slump onto nearshore ice. This process is particularly effective in high Arctic regions where Pleistocene glacial tills are emerging above the sea level as a result of postglacial isostatic rebound.[9] Coastal cliffs are receding in many parts of Arctic Canada and along the northern coast of Alaska (Reimnitz and Barnes, 1975, 1987). This means that considerable quantities of sediment are being removed from these coastal zones by, among other processes, sea ice rafting.

On the shallow shelves, a process called 'suspension freezing' can lead to the incorporation of sediment into sea ice (Reimnitz et al., 1992, 1993b). For this to occur, a storm is needed to agitate the water column with enough force to swirl loose particles from the sea floor and to lead to their suspension in the water. As the water freezes as a result of very low air temperatures, ice crystals form and, as they float to the surface, entrap some of the suspended sediment particles. The results of this process are sea ice floes that contain fine grained and somewhat sorted sediment at their underside. This positioning is extremely important for the distribution of these sediments in the ocean. Because melting removes some ice from the surface and winter freezing adds ice to the underside, the debris becomes sealed into the ice and gradually moves up as more and more ice is both melted from the top and added underneath. If drifting poleward, these sediments can survive several melting seasons without being lost. This process has important consequences for the interpretation of ice rafted sediments. But we will elaborate on this subject later.

At the present, most sea ice in the Arctic Ocean contains sediment at the surface, and in melt ponds where the sediment is concentrated in tiny, U-shaped holes called cryoconites. Sediment also occurs as irregularly shaped inclusions randomly oriented in the ice (personal observations by the author in the summers of 1989 with RV Polarstern, and in 1992 with RV Polar Star). In addition to clearly visible sediment inclusions, supposedly 'clean' sea ice without any visible sediment still contained small amounts of mineral and biotic material (Wollenburg and Pfirman, 1991). The sea ice sediments are mostly fine grained, in the silt and clay grain size range (Wollenburg, 1993; Reimnitz et al., 1993; Eicken et al., 1994), but coarser, sandy and gravelly sediment loads can also be found occasionally.

That ice contains sediment in various positions, as well as some of the processes that lead to sediment incorporation, has been known for a long time. Detailed and very lively descriptions and eye-witness accounts of the above-mentioned processes of sea ice sediment entrainment are given by Washburn (1947, pp. 80–84). Here are a few excerpts:

> 'As the spring and summer thaws proceed, land slips occur, and earth, gravel, and avalanches of stones come thundering down on the ice foot, there to remain until it breaks off from the coast, and floats out to sea with its raft-like load of land debris' (observation made in Greenland by Brown, 1875).

> 'In places rivers carry large amounts of dirty ice into the sea (Wrangell, 1840) or carry sediment out over sea ice (Nares, 1878)'.

The action of wind is described by Bell (1886):

> 'During the winter he (Mr. Skinner) found that the strong winds carried considerable quantities of dust and angular fragments of rock from the high cliffs and steep and loose taluses on either side of Nachuk Inlet, out upon the fixed ice, and when it broke up in July, this material was borne off to sea.'

The process of anchor ice formation also was known long before our time. Rae writes in 1850 that

> 'I had observed that large stones, some of them one or two tons weight, were making their appearance on the ice ... They could not have fallen from the shore as the beach was sloping nor had they been carried in by drift ice of the previous season. The only way that I could account for is this. At the commencement of winter the ice ... had become frozen to the stones lying on the bottom, and raised them up when the tide came in. The stones would get gradually inclosed in the ice as it grew thicker ... whilst by the process of evaporation, which goes on very rapidly in the spring, the upper surface was continually wasting away, so that in June and July there was little of the first formed ice remaining, and thus the stones which at first were on the under surface of the ice appeared on top.'

Finally, Washburn concluded that 'observations of debris in process of being carried by ice, and observations of the results indicate that ice rafting is a process of considerable quantitative importance ...'.

The sediment concentrations in sea ice are seasonally and regionally highly variable. They range over more than 6 orders of magnitude from values near zero to as much as 56,944 mg/l ice in the eastern and central Arctic Ocean (Nürnberg et al., 1994), and up to 920 mg/l in the Beaufort Sea (Reimnitz et al., 1993). In the Laptev Sea, a marginal shelf sea of the Arctic Ocean which is considered a major 'ice factory', the sediment loads reach values of 600 mg/l (Eicken et al., 1994). It should be borne in mind that these values represent snapshots of particular years, during which the Arctic Ocean was visited by icebreaker expeditions. The obtained values for any particular sub-region of the Arctic Ocean can vary considerably from year to year. For example, Tucker et al. (1999) found maximum concentrations of up 2228 mg/l in the summer of 1994 in the same part of the western Arctic Ocean for which Reimnitz et al. (1993) reported only 920 mg/l for the year 1988.

All expeditions into the Arctic Ocean from 1984 on have unanimously concluded that (a) the aereal distribution of visibly sediment-laden sea ice is irregular with no discernible geographic trends, and (b) the concentration of sediment in the ice varies over more than 6 orders of magnitude and also from year to year.

More elaborate treatments of the subject of sediment inclusions in sea ice, in addition to those already mentioned above, can be found in papers published by, most notably, Erk Reimnitz, Walter Tucker, Debra Meese, Ingo Wollenburg, Stephanie Pfirman, Hajo Eicken, Dirk Dethleff and Dirk Nürnberg. Here is a selection:

Pfirman, S., Gascard, J.-C., Wollenburg, I., Mudie, P. and Abelmann, A. (1989). Particle-laden Eurasian Arctic sea ice: Observations from July and August 1987. *Polar Research*, 7, 59–66.

Pfirman, S., Manfred, L., Wollenburg, I. and Schlosser, P. (1990). Sea ice characteristics and the role of sediment inclusions in deep-sea deposition: Arctic–Antarctic comparisons. In: Bleil, U. and Thiede, J. (eds) *Geological history of the polar oceans: Arctic versus Antarctic*. Kluwer Academic Publishers, Dordrecht; 187–211.

Pfirman, S., Wollenburg, I., Goldschmidt, P., Bischof, J. and Lange, M. (1991). Sediment in Barents sea ice. In: G. Krause, J. Meincke and H.J. Schwartz (eds) *Reports on Polar Research, Scientific Cruise reports of Arctic Expeditions ARK VI/ 1–4 of RV 'Polarstern' in 1989*. Bremerhaven: Alfred Wegener Institute for Polar and Marine Research; 33–35.

Reimnitz, E., Barnes, P.W. and Weber, W.S. (1993). Particulate matter in pack ice of the Beaufort Gyre. *Journal of Glaciology*; 39 (131): 186–198.

Reimnitz, E., Marincovich, L., McCormick, M., and Briggs, W. M. (1992). Suspension freezing of bottom sediment and biota in the Northwest Passage and implications for Arctic Ocean sedimentation. *Canadian Journal of Earth Sciences*; 29. 693–703.

Reimnitz, E., McCormick, M., McDougall, K. and Brouwers, E. (1993). Sediment export by ice rafting from a coastal polynya, Arctic Alaska, U.S.A. *Arctic and Alpine Research*; 25 83–98.

Reimnitz, E., McCormick, M., Bischof, J., and Darby, D. (1998). Comparing sea-ice sediment load with Beaufort Sea shelf deposits: Is entrainment selective? *Journal of Sedimentary Research*, 68 (5): 777–787.

Reimnitz, E. and Barnes, P.W. (1975). Sea ice as a geologic agent on the Beaufort Sea shelf of Alaska. In: Reed, J.C. and Sater, J.E. (eds) *The Continental shelf of the Beaufort Sea*. The Arctic Institute of North America; 301–354.

Reimnitz, E. and Barnes, P.W. (1987). Sea-ice influence on Arctic coastal retreat. In: Kraus, N.C. (ed.) *Coastal Sediments '87*. Proceedings of a specialty conference on advances in understanding of coastal sediment processes. New Orleans, LA, Vol. II: 1578–1591.

4.3 ICEBERGS

Icebergs are pieces of ice that break off the front of glaciers which reach sea level. Depending on the physical condition of the parent ice mass (the glacier), icebergs come in many different shapes and sizes. The large Antarctic ice shelves produce tabular icebergs of sometimes gargantuan dimensions. Just recently, an iceberg more than 160 km (as reported in the *New York Times*, 28 September 1999) across broke off Antarctica. This single berg is so large that it is considered a threat to shipping. Such large icebergs are not uncommon around the Antarctic ice shelves. Antarctic icebergs are large enough that theoretically they could survive a drift towards the equatorial regions. In fact, it was once seriously considered to use Antarctic icebergs as a freshwater source for arid regions by towing them there (Hult and Ostrander, 1973). At present, icebergs drift clockwise around Antarctica and therefore do not interact much with the sub-Antarctic waters. Even during the Pleistocene glaciations, the circum-Antarctic belt of drifting icebergs was only slightly larger (wider) than it is today (Lutjeharms, 1985; Westall and Fenner, 1990).

Icebergs that calve off the northern hemisphere ice sheets and glaciers are smaller than the Antarctic ones. The largest calve off West Greenland and have keel depths of up to 400 m. Other sources of icebergs in the northern hemisphere are the Svalbard archipelago, Franz Josef Land, Zevernaya Zemlya, and some of the northern and eastern islands of the Canadian Arctic Archipelago. In the northern hemisphere, icebergs are frequent in the Barents and Kara seas and in the eastern, Eurasian part of the Arctic Ocean, whereas the western Arctic Ocean is almost iceberg barren. The icebergs drift with the general Arctic circulation (TPD) towards Fram Strait and then southward along the East Greenland continental margin with the East Greenland Current. Along this drift they are joined by icebergs from the fjords of East and North Greenland.

Icebergs can contain considerable sediment loads, and the debris can be very coarse, up to boulder size (colour section, plates 1–15). Where icebergs contain debris depends on the morphology of the iceberg source. If the icebergs break off tidewater or valley glaciers, as for example in Spitsbergen, East Greenland or northern Ellesmere Island, debris can fall from the mountain flanks on the ice surface. If the parent glacier is crevassed, this debris can fall into cracks and then be in an englacial position. However, most glaciers have a debris-rich basal zone consisting of bedrock fragments that are being incorporated into the ice by mechanical crushing of the glacier as it flows across its bed (for more on this subject, see Drewry, 1986). These ice masses produce icebergs with debris at their underside. This type of debris incorporation was directly observed during a cruise into the Barents Sea in 1989. Relatively small, overturned icebergs that broke off the Austfonna Ice Cap, Nordaustlandet, Svalbard, contained sediment and red granite boulders at their underside. This discovery proved to be a boon for Norwegian geologists who were able to determine the bedrock geology underneath this ice cap. Until that cruise, the bedrock composition was completely unknown because the entire area was covered by more than 100 m thick ice which made the bedrock surface unaccessible. Thus, icebergs can even serve as bedrock indicators if the bedrock composition cannot be determined otherwise.

Another mode of particle incorporation into an iceberg was studied on the same cruise. An iceberg that had run aground in the Barents Sea was sampled by helicopter, and an ice core as well as several chunks of glacier ice were recovered (Jensen et al., 1991). The iceberg contained a thin, dark band just beneath its surface. A piece of ice with debris from this band was melted and the content inspected under a microscope. The band contained well-sorted, fine-grained (less than 0.5 mm) sand with grains of (in order of abundance) quartz, red k-spar, garnet, amphibole and biotite. From these properties it was concluded that (a) the sand was eolian in nature and was blown onto the parent glacier's surface by winds from nearby rock exposures, (b) that the parent glacier must be at least 100 m thick (based on the iceberg's dimensions), and (c) that the rocks that form the glacier's environment must be of a garnet-amphibolite and/or garnet-mica-schist type. This placed the origin of the iceberg into the fjords of northwestern Nordaustlandet, Svalbard (Pfirman et al., 1991). The iceberg had thus drifted at least 200 km.

As icebergs drift, they lose mass by melting, and the weight distribution changes. As a result, icebergs can roll over (which can lead to sudden loss of exposed debris), or they simply change shape by melting. A shape that often develops by these processes is the 'triple-crown', an iceberg that has vertical columns of remnant ice near the sides and a depression, often with a melt pond, in the centre. Sediment contained in an iceberg is primarily lost by melting. If ice with sediment melts, the sediment is either washed off the iceberg, or it concentrates in melt ponds. Eventually, all sediment will be lost and drop to the sea floor during the iceberg's life span.

In order to correctly identify the iceberg's drift tracks, the following considerations are important.

(1) The release of debris from an iceberg is a highly irregular, episodic process. Sediment is being lost on one or several point sources along the drift track, and not continuously. Thus, the lithologically interpretable trace of one single iceberg is discontinuous.
(2) After being washed off the surface, the sediment–water mixture forms a turbid suspension cloud in open water. Out of this cloud, the larger and heavier particles (sand, gravel and boulders) quickly drop to the sea floor on a nearly vertical path. By contrast, the finer-grained particles (clay, silt and fine sand) sink at lower velocities. One can reasonably expect that, because of the turbulent, tumbling motion of sand grains as they pass through the water column, the originally small-scale sediment patch disperses over a broad area upon arrival at the sea floor. This way, the relatively smaller grains are likely to produce a wide-diameter, thin, two dimensional, single-grain layer with a coarse spot in the centre. But the clay and silt-sized particles can be carried by deep water currents over long distances, far away from the actual location of the iceberg. This is also the reason why the clay mineral composition is not a suitable tool to trace ice drift directions.

To illustrate these points, let us consider the case of our tabular, grounded Barents Sea iceberg. The iceberg travelled at least 200 km after it broke off its parent glacier but had not yet lost the eolian sand from beneath its surface by the time we visited it. The further fate of the debris in this iceberg could have been like this: since the iceberg was visited in the middle of summer, with at least one more month of high solar insolation and melting to come, it is very likely that the surface of the iceberg would have experienced some more melting, and thus the sediment inclusion would appear at the surface, probably as a watery solution. If a slope towards the side existed, the water and the sand grains should have flowed down the side of the iceberg and would have hit either the surface of the pack ice or fallen into open water. If it hit the pack ice surface, it would have travelled with the West Spitsbergen Current into the northern Norwegian Sea and melted there, producing a thin layer of ice rafted sand grains with a provenance in northwestern Nordaustlandet. If the sand grains dropped into open water, the thin layer would be in the Barents Sea. As this example illustrates, the transport path of individual, ice rafted sand grains can be extraordinarily complex. First, our grains were blown by winds on the glacier surface. Then, they resided near the ice surface for a while as more snow accumulated and the ice thickness grew. Next, the iceberg calved off and drifted southeastwards to the site where it was sampled. Judging from the ice drift direction and the 100% sea ice cover, the further, speculative, fate of the sand grains was to be washed onto the sea ice surface, dropping somewhere west of Spitsbergen: a multiple-step, convoluted drift path of more than 700 km length.

In real nature, of course, not just one but many icebergs, or iceberg and sea ice combinations, travel along this same route and drop debris along the way. Over the course of sufficient time, this leads to a continuous trail of source-specific ice rafted lithologies on the sea floor that can be traced to their continental origin very much like the sand grains in the tabular iceberg.

The presence of icebergs has important implications for the physical properties (temperature, salinity and density) of the upper ocean. The melting of icebergs consumes heat from the atmosphere and the upper ocean and adds freshwater with a temperature close to the freezing point to the upper ocean. The combined effects are colder and lower saline surface waters. The freshening of the surface waters makes them specifically lighter and more sensitive to freezing in the winter. Icebergs and cold surface waters such as the East Greenland Current mutually stabilise each other. While cold surface waters facilitate long iceberg drift tracks, the melting of icebergs, in turn, helps maintain cold surface conditions.

The ultimate product of ice rafting, whether sea ice or iceberg rafting or a combination of both, is glaciomarine sediment. It is characterised by the presence of coarse (generally >63 μm) rock and mineral fragments in variable quantities. These coarse, terrigenous components cannot be found in distal oceans that have always been ice-free. In the high latitudes, the alternating glacial and interglacial climates have produced layers of coarse, glaciomarine sediment, and intercalated layers of fine-grained, mostly biogenic, interglacial sediment.

For more on icebergs, the reader is referred to:

Dowdeswell, J.A., Maslin, M.A., Andrews, J.T. and McCave, I.N. (1995). Iceberg production, debris rafting, and the extent and thickness of Heinrich layers (H-1, H-2) in North Atlantic sediments. *Geology*; 23 (4): 301–304.

Dowdeswell, J.A. and Dowdeswell, E.K. (1989). Debris in icebergs and rates of glaci-marine sedimentation: observations from spitsbergen and a simple model. *Journal of Geology*; 97: 221–231.

Dowdeswell, J.A., Whittington, R.J. and Hodgkins, R. (1992). The sizes, frequencies, and freeboards of East Greenland icebergs observed using ship radar and sextant. *Journal of Geophysical Research*; 97: 3515–3528.

Dowdeswell, J.A., Villinger, H., Whittington, R.J. and Marienfeld, P. (1993). Iceberg scouring in Scoresby Sund and on the East Greenland continental shelf. Marine Geology; 111: 37–53.

Dowdeswell, Julian A. (1995). Deep Pleistocene iceberg plowmarks on the Yermak Plateau: Sidescan and 3.5 khz evidence for thick calving ice fronts and a possible marine ice sheet in the Arctic Ocean. Comment, *Geology*, May 1995: 476–477.

Vogt, P.R., Crane, K. and Sundvor, E. (1994). Deep Pleistocene iceberg plowmarks on the Yermak Plateau: Sidescan and 3.5khz evidence for thick calving ice fronts and a possible marine ice sheet in the Arctic Ocean. *Geology*, 22: 403–406.

Vogt, P.R., Crane, K. and Sundvor, E. (1995). Deep Pleistocene iceberg plowmarks on the Yermak Plateau: Sidescan and 3.5khz evidence for thick calving ice fronts and a possible marine ice sheet in the Arctic Ocean. Reply, Geology, May 1995: 477–478. (Both papers by Vogt et al. are part of an interesting controversy: are iceberg plowmarks in water depths of up to 1000 m really evidence for similarly thick ice sheets? Or could they be produced by thinner, up to 400 m thick, ice sheets, when large tabular icebergs capsize and roll over, obtaining keel depths that are much larger than the thickness of the parent ice mass, as suggested by Julian Dowdeswell in a comment?)

5

The Norwegian Greenland Sea ice rafted record

The first example of ice rafting comes from the Norwegian Greenland Sea (NGS). The NGS is a fascinating study area from a paleoclimatological viewpoint owing to a large climate contrast between its eastern and western sides, and because it has experienced some of the most drastic climate changes during the Pleistocene. The NGS is an approximately rectangular-shaped northern extension of the North Atlantic Ocean (Fig. 5.1). It connects the very cold, ice covered Arctic waters with the warmer, ice-free North Atlantic waters. At the present, the NGS is clearly divided into two parts: the eastern, ice-free half with northward flowing waters of North Atlantic origin, and the western half with cold, southward moving ice covered waters of Arctic Ocean origin (Johannessen et al., 1994). The thermal gradient between these two parts is large. On the eastern side, the climate is relatively warm and hospitable, while on the western side (Greenland) it is brutally cold almost all year round. During the Pleistocene glaciations, the NGS was surrounded by large ice sheets which in addition to Greenland, covered Iceland, Svalbard, the Barents shelf and Scandinavia (Boulton et al., 1985). Therefore it was expected that the NGS sediment record would reveal some of the most drastic glacial to interglacial climate changes on Earth.

The purpose of my study of IRD and dropstones (ice rafted rock and mineral fragments large enough to be visible in the opened core section) in the NGS was to obtain a regional overview of the ice rafted lithologies, connect them as far as possible to their most likely continental and shelf sources, and reconstruct the Pleistocene ice drift and surface current directions. At the time the project started in 1986, not much was known about the regional and vertical distribution of dropstones in this area. The only prior information came from Wohlfeil, who had analysed dredged dropstones from the Iceland-Faroer Ridge, a relatively shallow (in parts less than 400 m water depth) sill between the Norwegian Greenland Iceland seas to the north and the North Atlantic to the south. He found that most of the rocks in his study area came from southern Scandinavia and the Baltic region and concluded that they were transported by icebergs that calved off the front of the Late Weichselian (in US: Wisconsinan) Fennoscandian Ice Sheet. To a similar conclusion also came Ruddiman (1977), who proposed that a circulation in form of a

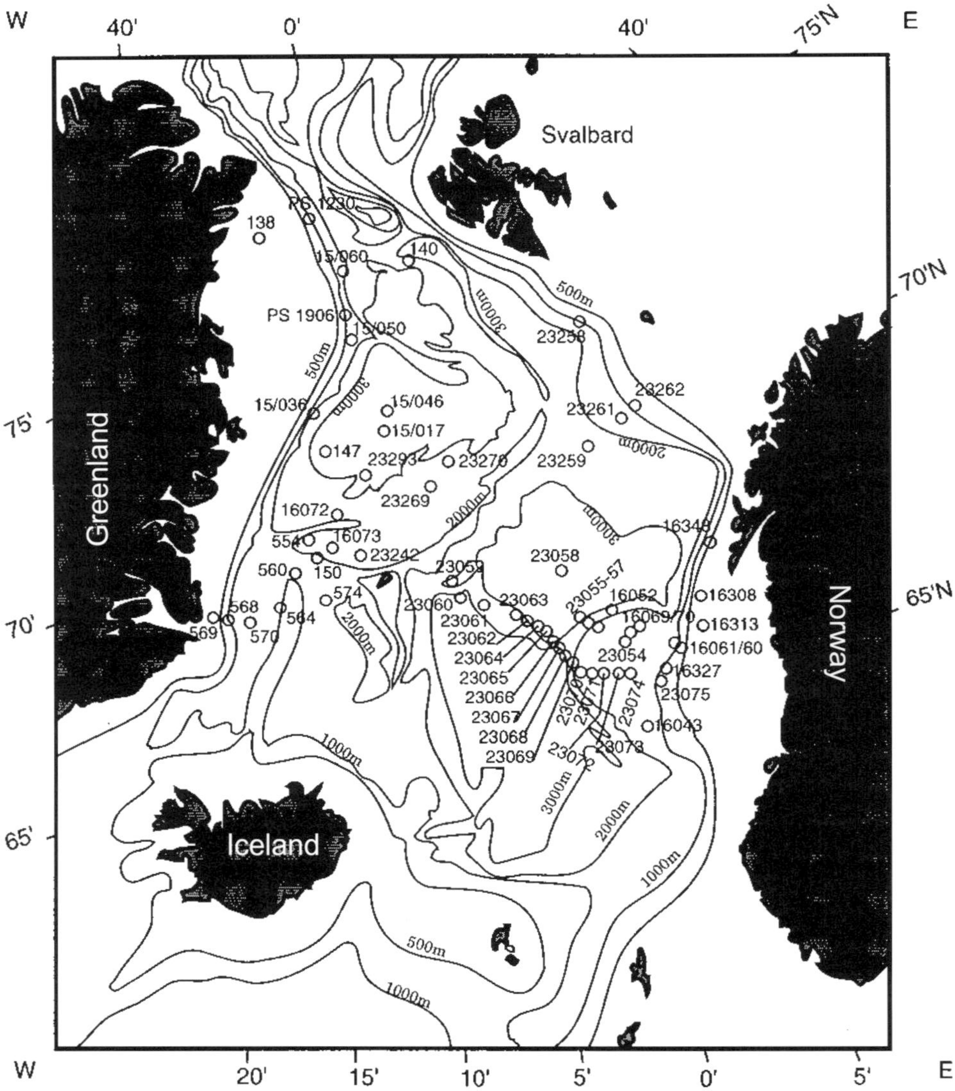

Fig. 5.1. Map of the Norwegian Greenland Sea with all core locations.

counterclockwise loop in the Norwegian Sea would best explain the distribution of ice rafted sand grains in the North Atlantic. However, he did not analyse cores from the Norwegian Sea. Some information also was available from ODP leg 104, Vøring-Plateau, located to the west of Norway between 65 and 70 degrees North. In cores taken from three boreholes, Krissek (1989) performed a semi-quantitative analysis of ice rafted debris and noted the frequent occurrence of chalk, a rock type that does not crop out anywhere near the core site. He concluded that ice from the south must have drifted to the Vøring-Plateau at some time and dropped these fragments along the way.

The objective of my Norwegian Greenland Sea work was to identify the types of ice rafted rock and mineral fragments that occur in deep-sea sediments, to link them to their most likely continental sources, and to reconstruct the past ice drift directions in chronologic order. The sediment cores available for study were large (50 × 50 × 50 cm) box cores and long gravity cores with a quadratic cross section of 30 × 30 cm ('Kastenlot' in German). The maximum penetration depth of the cores was 8 m, which translates into an age of up to 450,000 years for the oldest sediments. A typical Norwegian Sea average sedimentation rate is 1.7 cm/kyr for the distal ocean, increasing towards the continental margins. The box core sediments have maximum ages of up to 30,000 years. This time frame places the box core samples into the Late Weichselian to Holocene, while the longer gravity and kasten core samples represent three or more glacial–interglacial cycles (more on this specific subject in Henrich, 1990 and Henrich et al., 1989).

Only vague ideas existed about what we might find. It was speculated, for example, that most of the ice rafted lithologies originated from coastal parts of Norway (this turned out to be true), and that the surface currents that moved the icebergs through the Norwegian Sea may have changed with time (also true). In order to make paleo-current or paleo-ice drift reconstructions possible, it was hoped that the regional distribution of IRD would be uniform within certain sub-sections of the NGS, and thus be interpretable. Because of the erratic nature of iceberg rafting, dropping some stones here and some there, there was a need to demonstrate that these individual iceberg droppings had, over the course of hundreds or thousands of years, created layers with relatively uniform petrographic composition of distinct geographic origin. In other words, the content of the analysed layers does not contain just the debris of one single iceberg, which would make the sample rather unrepresentative, but instead is a collection of perhaps thousands of icebergs. These conditions, as we shall see, were met. The dropstone work was a part of comprehensive sedimentological, paleontological, geochemical and biological studies aimed at a better understanding of the climatic history of the Norwegian Greenland Sea under the umbrella of special research project 313.

5.1 NORWEGIAN SEA GLACIOMARINE SEDIMENTS AND METHODS OF INVESTIGATION

Sediments in the Norwegian and Greenland seas consist of alternating sections of light to medium brown mud with low to moderate quantities of ice rafted debris, and dark grey, dark olive grey, and black layers with high to very high IRD quantities. As a general rule, these coarse units contain few marine microfossils, while the brownish muds contain variable quantities of marine microfauna and -flora remains. The colour banding can be traced over several 100 km in the Norwegian Sea. Toward the west, the colour bands give way to a more uniform type of glaciomarine sediment in shades of grey. An excellent description of the sediments in the Norwegian Sea can be found in Henrich et al. (1989). An abbreviated description is presented below.

The youngest, Holocene, sediments at the top form a fine-grained surface layer with abundant foraminifera. This layer is 10–15 cm thick in the western and central Norwegian Sea and considerably thicker towards the Norwegian continental margin, where its thickness can exceed the penetration depth of the box cores (circa 50 cm). This was the case for box cores 23068–23071 (Fig. 5.1). The surface layer contains very few coarse, ice rafted rock fragments (only 1–20 lithic grains >0.5 mm per litre sediment), and was categorically separated from the underlying glaciomarine sediments. This was done to ensure that the Holocene samples contain only Holocene, post-glacial sediments and not remnants of the latest Pleistocene glacial sediments. Within these glacial sediments, three very coarse-grained, dropstone-rich units of up to 15 cm thickness are separated from each other by thin (on average 1 cm thin), relatively fine grained layers. In the western Norwegian Sea, the coarse layers are increasingly amalgamated together. This type of layering could be the Norwegian Sea analogue of the now-famous 'Heinrich Layers' in the North Atlantic.

The longer kastenlot sections contain at least three full glacial–interglacial cycles. Every glacial cycle begins (at the bottom) with a yellowish-brown, sandy mud with variable amounts of marine microfauna remains and low to moderate concentrations of ice rafted debris. The debris is randomly oriented and does not form coherent, distinct layers. X-ray photographs of this sediment type show that individual ice rafted rock fragments appear to be 'floating' without grain-to-grain contact in a fine-grained matrix. The glacial cycles before the Late Weichselian Glaciation (marine isotope stages 6, 8, 10, and 12) terminate at the top with dark grey, dark greenish-grey, or even black, sandy to gravelly mud. This sediment is firm to very firm, somewhat overconsolidated, and contains coarse IRD in maximum concentrations and in distinct layers. In contrast, the last glacial cycle (isotope stage 2) terminated with several layers of a similarly coarse, sandy to gravelly mud, but without the dark greenish-grey to black colouration. Every glacial cycle has a relatively thick, lower sediment section with moderate quantities of coarse IRD and a thinner top with maximum IRD content. It appears from this that relatively long periods of ice sheet growth on the continents and shelves around the NGS were followed by comparatively short and drastic terminations, during which large quantities of icebergs were ejected from the waning ice sheets. During the earlier ice-growth phase, only low to moderate numbers of icebergs reached the open ocean, whereas the end is a stepwise crescendo with maximum iceberg calving and rapid ice margin retreat.

5.1.1 Sampling

The sampling strategy and the analytical procedures are described in more detail below.[10] Here is the short version. The samples are horizontal slices of sediment that were cut from the open cores. The upper Holocene layer was split from the glaciomarine sequences below. In the Pleistocene section, the very coarse, IRD rich layers were separated from each other to ensure that the IRD results refer to the same stratigraphic levels. The samples were wet sieved through 0.5 mm screens.

5.1.2 Sample processing

The analytical procedure is described in detail below.[11] A random split of the >0.5

mm fraction was analysed quantitatively under a binocular microscope at 20 to 40 × magnification. Pieces of up to 10 cm in diameter were usually found abundantly in the Pleistocene sections and used for comparison with the smaller fragments and to verify the petrographic determinations. In addition to this, thin sections were made to accurately determine the IRD petrographies. In total, 114 different types of rock and mineral fragments were distinguished.

5.2 DROPSTONES IN THE NORWEGIAN SEA

Dropstones in the Norwegian Sea consist of highly diverse assemblages of crystalline and sedimentary rocks, quartz and feldspar grains. The size of the grains ranges from the arbitrarily selected cut-off value of 0.5 mm to blocks of up to 30 cm in diameter. The largest blocks were found in the Greenland Sea. The size of the truly largest dropstones may never be known, due to the limited diameter of the coring devices. But occasional coring failures, known as 'bananas' because of the shape of the core barrel after recovery, are thought to be the result of large stones that were hit by the barrel. The frequency of 'bananas' is thus an indirect measure of the presence of large stones in deep sea sediments. A 'typical' Norwegian Sea dropstone sample contains, in addition to the most abundant sand-sized grains, several pieces of larger rocks up to fist-size.

Virtually any rock type imaginable can be found in Norwegian Sea IRD. In fact, the content of most samples could serve as a veritable petrography lesson. The crystalline rocks are gneisses and granites in colours of white-clear, red, orange, pink and yellow; mafic rocks (gabbros, norites, amphibolites, serpentinites, etc.) of grey, black, and dark green; quartzites in a large variety of different colours including grey, white, green, orange, red, yellow and black; and phyllites and mica schists in grey and silver. Gneisses are the most abundant type of crystalline rocks, followed by (in decreasing order) light and dark coloured igneous rocks, and quartzites. Meta-argillites and volcanic rocks are generally less abundant. The gneisses display a pronounced banded and layered structure, with the bands occasionally wrapped around larger porphyroblasts of feldspar ('augengneiss'). The rocks are similar to erratic boulders in glacial tills from northern Germany, and to Norwegian rocks in the collection of the Geological Museum in Oslo. Lithologic resemblance to Scandinavian, and in particular Norwegian, rocks of course does not constitute proof of origination, because similar rocks also occur in the northern British Isles, in Greenland, and in parts of Svalbard. All these areas were, and in part still are, glaciated and thus could have produced icebergs with crystalline material drifting to the Norwegian Sea.

The sedimentary rocks are almost exclusively clastic sedimentary rocks. Most abundant are grey siltstones, followed by light grey to white sandstones, black shales, and brown siltstones. Red and green clastics and red, ochre and yellow clay ironstones also occur in smaller percentages. Carbonates are very rare and fall into two groups: (1) true limestones with a fine grained, diffuse matrix and microscopic fossils of algae, and (2) chalk, a rock that is entirely composed of coccolith shells.

Chalk always occurs alongside a medium to dark grey chert, called flint. While these rocks are generally rare (2% or less of the >500μm grains), they are nevertheless common among dropstone assemblages dominated by crystalline rocks. The significance of this lies in the fact that chalk and flint are only present in Late Cretaceous formations to the south of 59°30N (Hancock, 1984). Their presence thus unequivocally proves a southeastern derivation with respect to the Norwegian Sea core locations, and an ice drift direction north- and northwestward up along the Norwegian coast. The other carbonates (limestones and dolomites) are different from those in IRD assemblages in the Arctic Ocean (Clark et al., 1980; Bischof et al., 1996; Bischof and Darby, 1997). In the Arctic, carbonate IRD is macroscopically light brown to medium grey, under the binocular, white, brown, grey and yellow, more rarely black, orange, pink, red and green. The rocks are featureless, microcrystalline and devoid of microfossils. But the Norwegian Sea IRD carbonates are not recrystallised, more uniformly grey, and contain microfossils. Most frequent were tiny algal structures (detected in thin sections) and bryozoan rests. Thus, an Arctic provenance for the carbonates among IRD of the NGS is excluded.

The average IRD composition of Norwegian Sea sediments (grains >0.5 mm) is shown in Table 5.1. On average, crystalline rocks are most abundant with 30.9%, followed by clastic sedimentary rocks (27.5%), quartz grains (26.2%), and feldspar (12.5%). Crystalline rocks are composed of 9.9% light coloured gneiss, 5.3% mafic gneiss, 4.8% each quartzites and granites-syenites-tonalites-diorites, 3.1% volcanics, and 1.2% phyllites and mica schists. Limestones and dolomites are less than 1%, chalk and flint even less than 0.5%.

The range of the data variability is enormous. Of the 19 listed component groups, only two – clastic sedimentary and crystalline rocks – are invariably present in every sample, regardless of their stratigraphic position. All other components vary between 0% and multiples of their average values. In most cases, the maximum concentrations are 2–5 times higher than the average value. Exceptionally high variability is noted for coal, which can be 46 times as abundant as its average, and volcanic rocks (12 times the average).

In comparing the Pleistocene, which here contains the time period from oxygen isotope stage 12 (approximately 450,000 years bp) to the end of the Late Weichselian, approximately 10,000 years bp, with the Holocene (post 10,000 ybp), the Pleistocene has higher concentrations of clastics and feldspar, but is lower in volcanic rocks and coal. The differences exist for averages as well as for maximum concentrations. Limestones and dolomites show a stronger affinity to the Pleistocene section, the chalk and flint delivery is unchanged, but other types of chert such as spiculites from central Spitsbergen (Siedlecka, 1970) are more frequent during the Holocene. Crystalline rocks combined are slightly more abundant in the Holocene, which contains at least 14% crystalline rock fragments. Both light and dark coloured intrusives are equally abundant in both time periods. Light and dark coloured gneisses are equally abundant in their average values, but the maximum values are higher in the Pleistocene. The quartzite average is higher in the Holocene, but the maximum values are equal.

Table 5.1. Average composition of Norwegian Sea ice rafted debris

	All samples			Pleistocene			Holocene		
n samples	166			114			52		
grains counted	139728			115648			24080		
Percentages	Average	Minimum	Maximum	Average	Minimum	Maximum	Average	Minimum	Maximum
clastics	27.54	0.72	97.41	30.23	0.72	97.41	21.65	1.87	59.28
limestones and dolomites	0.95	0.00	8.30	1.00	0.00	8.30	0.82	0.00	4.79
chalk and flint	0.41	0.00	2.72	0.47	0.00	2.72	0.29	0.00	2.10
cherts (including spiculites)	0.30	0.00	6.67	0.21	0.00	1.67	0.50	0.00	6.67
crystalline rocks	30.86	0.26	67.31	28.47	0.26	61.48	36.09	13.99	67.31
granites, tonalites, diorites, etc	4.78	0.00	20.58	4.52	0.00	18.57	5.35	0.00	20.58
gabbros, norites, other mafics	0.95	0.00	5.34	0.96	0.00	5.34	0.93	0.00	4.88
light coloured gneiss	9.91	0.00	28.53	10.12	0.00	28.53	9.44	0.00	21.47
mafic gneiss	5.30	0.00	20.76	5.55	0.00	20.76	4.74	0.00	14.02
quartzites	4.83	0.00	15.49	3.89	0.00	15.37	6.90	0.00	15.49
phyllites and mica schists	1.24	0.00	9.15	1.27	0.00	5.47	1.18	0.00	9.15
volcanic rocks	3.10	0.00	50.66	1.08	0.00	9.65	7.54	0.00	50.66
quartz	26.18	0.00	53.42	24.93	0.00	53.42	28.91	3.13	53.3
feldspar	12.46	0.00	39.49	13.82	0.00	39.49	9.49	0.00	20.31
mafic minerals	0.36	0.00	6.11	0.18	0.00	1.36	0.75	0.00	6.11
coal	0.94	0.00	45.83	0.69	0.00	20.46	1.50	0.00	45.8
olivine	0.17	0.00	6.11	0.02	0.00	0.37	0.52	0.00	6.11
garnet	0.06	0.00	1.79	0.06	0.00	1.09	0.06	0.00	1.79
biotite	0.04	0.00	1.18	0.02	0.00	0.31	0.08	0.00	1.18

IRD (>0.5 mm) composition (average percentages, minima and maxima) of all samples with 23000 numbers from the Norwegian Sea, sector from 74 deg North and 5 deg West to 66 deg North and 6 deg East. Listed are all samples (left), Pleistocene samples (middle), and Holocene (right). The Pleistocene here refers to samples that go back to isotope stage 12, approximately 450,000 years before the present. The average values are based on every sample's individual percentages, not on the combined grain counts. To calculate the average percentage of clastics for all samples, for example, all 166 percent values were added and divided by 166. This method eliminates the effect of different grain counts.

5.3 COMPOSITION, LITHOLOGIC DIVERSITY AND SOURCES OF NORWEGIAN SEA IRD

The Norwegian Sea IRD is petrographically and mineralogically extremely diverse. No two samples are exactly alike. In an attempt to assess the full range of different rock and mineral varieties in a sample, one large volume sample (23063, 23–33 cm depth, ca. 18 - 14.4ka) was completely disassembled into as many different lithologies as this author can distinguish. In total, there were 363 different rock and mineral varieties. Most lithologies were crystalline rocks, different kinds of granites, gneisses, gabbros, amphibolites, quartzites etc., but there were also more than 60 different types of clastic sedimentary rocks and over 20 different carbonates. There were always several pieces of each discerned rock type. In the case of the clastics, up to several hundred grains of the same lithology could be found. Sample 23063, 23–33 cm was chosen for this experiment because it is very representative for Late Pleistocene Norwegian Sea ice rafted debris. The composition of Norwegian Sea IRD is extremely complex and mixed together with material from many different sources.

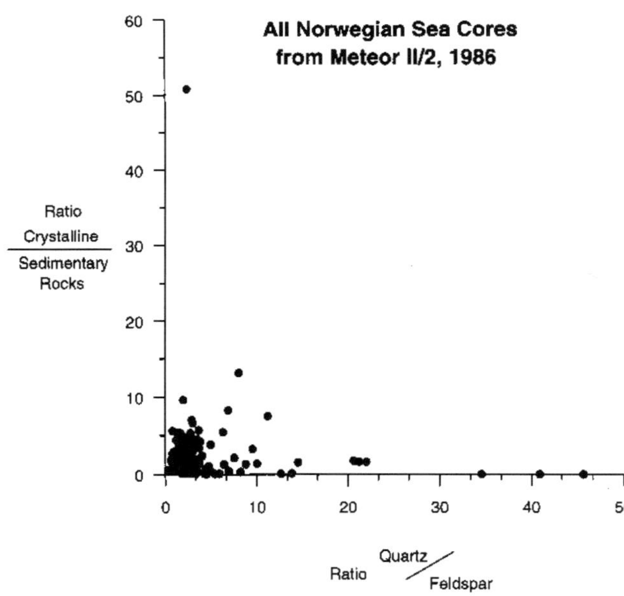

Fig. 5.2. X–Y plot of the four major IRD lithologies in the Norwegian Sea. X-axis: ratio of quartz to feldspar; y-axis: ratio of crystalline to sedimentary rocks. Note the dense cluster of data points in the lower left corner of the diagram. Most samples have both ratios in the range of 0–15, and more closely in the range of 0–5. One sample to the top left (23074, 545–565 cm, 47–45ka) has an exceptionally high crystalline to sedimentary rock ratio of >50, while another set of samples spreads along the x-axis to values of up to 46 for the ratio of quartz to feldspar. These latter samples contain very high quantities of clastic sedimentary rocks from the Barents Shelf.

In order for the reader to judge how reliable and trustworthy the source determinations, and therefore the resulting paleoceanographic conclusions, are, it is necessary to understand which kind of reasoning was used to determine the IRD sources. In a sense, the following is a description of the scientific philosophy that I have adopted to identify the provenance of ice rafted debris. Several lines of reasoning, both inductive and deductive, as well as the application of simple sedimentologic criteria, were used to identify the major IRD sources. Inductive reasoning, straightforward determination of the source, is possible in some cases because of the presence of unique lithologies from known places of origin in the IRD. For example, there are rock types such as rhomb porphyry, a volcanic rock only found in the Oslo graben, southern Norway (Oftedahl, 1952, 1981), and chalk and flint, rock types that occur underneath the North Sea and in the southern Baltic region, including southern Sweden, Denmark and northern Germany (Hancock, 1984). Another source indicator is spiculite, a chert with sponge spicules whose only known occurrence is in central Spitsbergen (Siedlecka, 1970). Spiculites were found in Late Pleistocene sediments of five cores from the Norwegian Sea (box cores 23055, 23056, 23059, 23065 and 23068; Fig. 5.1). Another source indicator is a particular fox-red biotite-quartz schist that was identified in a thin section of core 23059, 28–38 cm core depth. This rock is truly unique, was detected only once in the IRD, and comes from northwest Spitsbergen (Hjelle and Ohta, 1974). Inductive reasoning allows us to accurately pinpoint the sources of a few IRD fragments, and to reconstruct the actual ice drift directions. In the case of the spiculites, undeniably a southward drift must have operated. However, the quantities of most of the aforementioned lithologies among IRD are very low, which somewhat diminishes their value as a quantitative direction indicator. As opposed to chalk and flint, which are common in almost all glaciomarine samples from the Norwegian Sea, and which thus document an important ice drift direction, spiculites, rhomb porphyry and the fox-red biotite rock are very rare, and thus identify ice drift directions that were more exceptional.

Deductive reasoning is used to stepwise eliminate unlikely source areas. For example: Norway is composed of 99% (in percent surface exposure) metamorphic and igneous rocks. Only 1% of the Norwegian surface is composed of unmetamorphic sediments (Oftedahl, 1980). Therefore, if Norway had been the sole source of IRD for the Norwegian Sea, the resulting IRD composition >500 µm should consist of 99% crystalline lithologies (rock fragments and quartz/feldspar grains), and only 1% sedimentary rocks. This amounts to a ratio of close to 100 between crystalline and sedimentary rocks. But Fig. 5.2 shows that such a ratio is never reached. The highest ratio of 51 is from an exceptional sample from the western continental slope of Norway, which probably contains the purest, almost entirely Norwegian-derived ice rafted debris. But for most samples, the ratio between crystalline and sedimentary rocks is less than 10. This means that most samples contain a major contribution from Norway, but also lesser amounts of IRD from other sources composed of sedimentary rocks. The geographically closest candidates during the Late Weichselian glaciation are the northern British isles, the Baltic region (southern and southeastern Sweden, Denmark and northern Germany), central East Greenland, Svalbard and the adjacent Barents shelf, and the continental shelf off Norway.

The third line of reasoning is the application of simple sedimentological criteria. Simply put, the relative percentage of a certain IRD lithology must increase towards its origin. This method was used to identify the western Barents shelf as a major source of clastic sedimentary rocks during the Late Weichselian (Bischof, 1994). And the fact that most, if not all, crystalline lithologies in the Norwegian Sea IRD are derived from Norway is concluded from their increasing proportion towards the Norwegian continental margin (Fig. 5.3). The percentage of crystalline rocks among IRD increases by 57% from the central Norwegian Greenland Sea towards the Norwegian side (Fig. 5.3a). If the percentages are converted into percentages of all rock fragments (without mineral grains), the increase is even more drastic, 110% (Fig. 5.3b). Conversely, clastic sedimentary rocks increase towards the west, away from the Norwegian continental margin (Figs 5.3c, d), indicating that their main sources are located towards the west and north of the Norwegian Sea core locations.

Therefore, a major source of crystalline rocks was Norway, and possible sources for the sedimentary rocks were southern Sweden, Denmark, the outer Norwegian Shelf, central East Greenland, or Svalbard and the Barents Shelf. Of those, central East Greenland (Jameson Land) must be eliminated, because the palynomorph assemblages from the mostly Jurassic formations there (Lund and Pedersen, 1985) do not match those from clastic sedimentary dropstones from the Norwegian Sea (Bischof et al., 1997). North Greenland remains a likely source owing to the presence of Paleozoic acritarchs in ice rafted black shales from the Greenland and western Norwegian Sea (cores 23059 and 23060, Bischof et al., 1997). The Norwegian shelf is an unlikely source of clastics during the Weichselian Glaciation except for the last glacial maximum, 23–18ka, because that was the only time during which Fennoscandian glaciers reached the outer shelf (Mangerud, 1991a, b). On the Norwegian continental shelf, crystalline rocks are exposed on the inner shelf to a distance of circa 60 km offshore (Oftedahl, 1980). Only there does the bedrock composition change to sedimentary rocks. In fact, we could only once positively identify a fragment of a soft, greenish-grey Albian claystone as coming from the bottom of the Norwegian shelf by means of its palynomorph content (Bischof et al., 1997).

Additional information about the prevailing drift directions can be extracted from volcanic rocks. In the western Norwegian Sea, the volcanic rocks are black, dark grey and dark green basalts. The rocks have a fresh, unaltered appearance and are porous with rests of volcanic glass in intergranular spaces. These basalts are invariably associated with bottle-green olivine and porous obsidian in colours of black and dark brown, more rarely of dark green. The maximum percentage of these components is 22% of the > 500 μm fraction in the vicinity of Jan Mayen, decreasing steeply towards the east. In the eastern Norwegian Sea, volcanic rocks are present in small quantities, but they are petrographically different from those of Jan Mayen. They have lighter colours, reddish-brown to light grey, and are strongly altered. The pronounced alteration features suggest that they are 'old' volcanics, Paleozoic or Precambrian. Their most likely origin is in southern Norway.

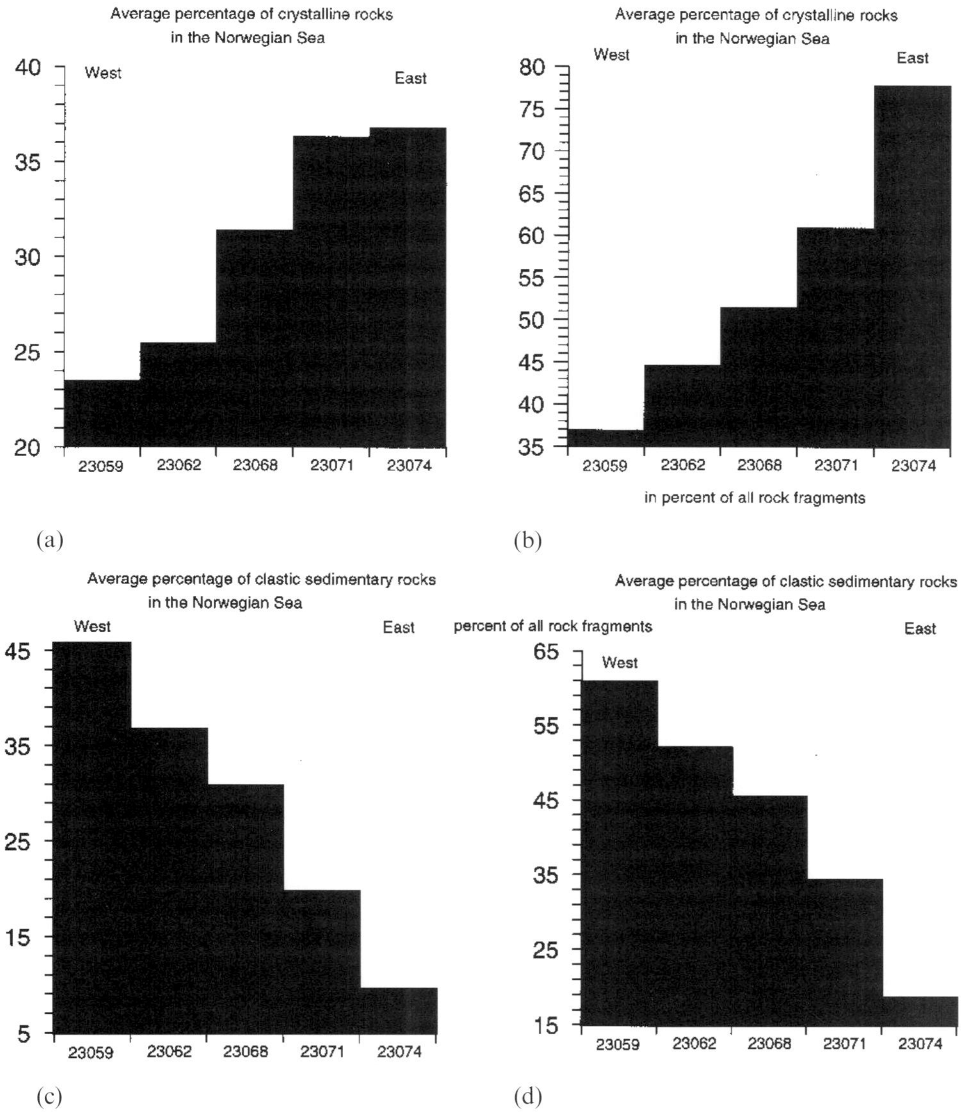

Fig. 5.3. Average percentage of crystalline rocks in the Norwegian Sea. Database: all combined GKG and KAL records, arranged geographically from west to east. Note the unambiguous increase of the crystallines towards the east (Norwegian) side of the Norwegian Sea. Core 23059 is in the western Norwegian Sea at 3° West, and core 23074 is at the Norwegian continental slope at 4°54 East. The eastward increase in the crystalline rock proportion is 56.9% for the entire >500 m fraction (a), and 111% for their increase among the rock fragments (b: excluding quartz, feldspar, and other mineral grains); (c) percentages of clastic sedimentary rocks among IRD >500 μm along the same transect; (d) clastic sedimentary rocks in percent of all rock fragments.

5.3.1 The general Pleistocene circulation ...

The generalised circulation necessary to produce this IRD distribution pattern is a counterclockwise gyre in the Norwegian Sea (Fig. 5.4). Flow in the eastern Norwegian Sea is generally to the north, in the northern Norwegian Sea it is westward, and in the Greenland Sea southward. This flow regime would permit crystalline lithologies from western and southern Norway, and chalk and flint from the Baltic region, to reach the eastern Norwegian Sea, while the western Norwegian Sea (or the central Norwegian Greenland Sea) received a blend from more northern sources, northern Norway, Svalbard and the Barents Sea, and perhaps also Siberia.

This circulation is fundamentally different from the present situation: it represents a southward shift of a counterclockwise gyre, which at present is located in the Fram Strait (circa 80° North), to a position at approximately 67° North in the eastern Norwegian Sea. This gyre is a very important climatic feature of the Fram Strait today. In it, the northward moving, normally-saline waters of the Norwegian Current cool to a temperature at which they become specifically heavier than the surrounding colder and less saline Arctic waters. They sink and slide underneath the Arctic surface waters and then spread in water depths below 400 m across the entire Arctic Ocean (Carmack et al., 1997). This same situation existed during the ice ages at the more southern location in the Norwegian Sea, but with a twist: because of the amount of freshwater that was stored in the large continental ice sheets, the average salinity of the global ocean was higher than today, perhaps at a level of 37 per mil. As a result, the northward moving current, which in Fig. 5.4 is shown in the lower right corner, had a higher initial salinity and would be heavier than the Norwegian Greenland Sea surface waters at that time, which were less saline and lighter than today owing to the icebergs and meltwater from the ice sheets. The Pleistocene precursor of the Norwegian current would sink and slide underneath the Norwegian Greenland Sea surface waters. It would thus not block the west- and southward motion of the surface waters and icebergs depicted in Fig. 5.4 for the area to the north and west of the gyre. In fact, the most likely northward flow of these Pleistocene waters at intermediate depth would provide some of the kinetic energy to drive the surface circulation.

But if significant amounts of freshwater, for example from melting icebergs or direct meltwater discharge from the continental ice sheets, mixed with the approaching North Atlantic waters, both the salinity and temperature of these waters would drop, possibly to the level of the waters in the Norwegian Greenland Sea. The density difference would disappear, and instead of flow at intermediate, sub-surface depths, the northward flow would be at the surface. Under this scenario, the northward drift of ice could continue into the Fram Strait and even reach the Arctic Ocean. And that is precisely what a collegue of mine (Spielhagen, 1990) found in a sediment core from the Fram Strait: chalk fragments from the Baltic area in sediments from the coldest part of the last glaciation, the last glacial maximum between 23 and 18 kyr. Even though these chalk fragments are exceedingly rare (only two pieces in all), they nevertheless demonstrate that northward moving currents with ice have reached this far northern location during the Late Weichselian Glaciation.

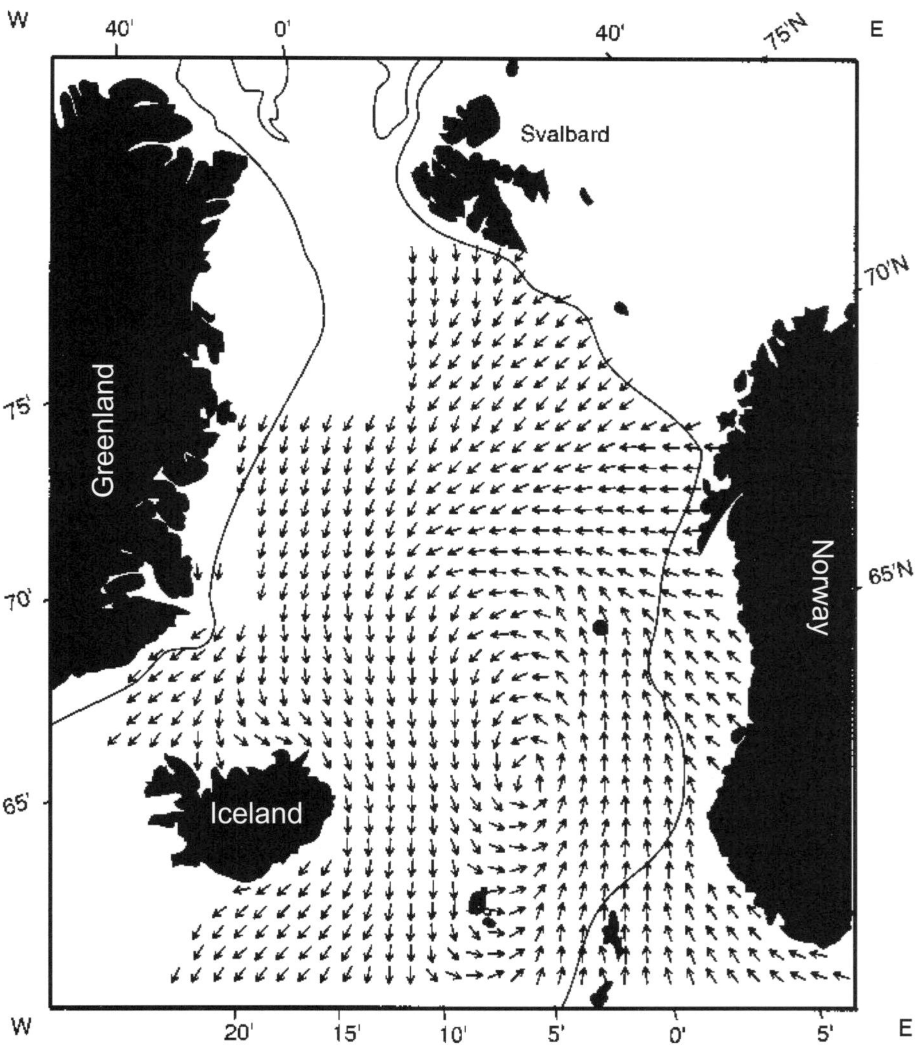

Fig. 5.4. Map showing the general circulation in the Norwegian Greenland Sea during the Pleistocene. This circulation produces a southern and western Norwegian and Baltic type of IRD composition in the eastern Norwegian Sea, while the western part is increasingly affected by rocks of more northern provenance, northern Norway, Svalbard, the circum-Arctic land areas and northeast Greenland. The outlines of ice shields are not shown, because they have shifted constantly, and are partly unknown for older, pre-Weichselian glaciations. This graphic shows an average situation that best approximates the determined IRD composition. At any given time period, or 'time-slice', within the Pleistocene, the ice drift deviated from this picture. But the deviations, as we shall see, were mostly variations of this theme, and not completely different situations (black dot: KAL 23074).

Figure 5.4, therefore, depicts a somewhat artificial average Pleistocene situation, from which the actual drift at any given time period deviated. These deviations, as the following discussion of the IRD composition will show, are subtle variations of the same theme: at times, icebergs took a more westward instead of northwestward drift track from the Norwegian side; at other times, the drift was pushed eastward from the Greenland side; at yet other times, the drift was more southward from the Barents ice sheet than shown in Fig. 5.4.

5.3.2 ... And its driving forces

The driving forces for these changes are in part atmospheric and in part oceanic. When ice sheets retreat, they introduce huge quantities of glacial meltwater and icebergs into the ocean, thereby increasing the relative sea surface height close to the ice margin. The water has to flow perpendicularly away from this surplus to even out the potential difference and would be deflected to the right because of the effect of the earth's rotation (Coriolis acceleration). If all ice sheets around the Norwegian Greenland Sea contributed at the same time, a counterclockwise surface circulation and ice drift very much like the one in Fig. 5.4 would develop. But the atmosphere could accomplish the same: large ice sheets form persistent atmospheric high pressure systems, which at the height of the ice ages were evenly distributed around the Norwegian Greenland Sea. The air flows from high to low pressure, towards the centre of the Norwegian Greenland Sea, and transmits energy via friction onto the ice-covered ocean. Again, the effect of the Earth's rotation would induce a deflection to the right and the result would be a similar counterclockwise gyre. Whether the ice drift and surface circulation at different times is dominated more by west-, east- or southward directions depends on the differential strength of the atmospheric pressure over, and the meltwater and iceberg discharge from, each of the surrounding ice sheets.

5.3.3 The mixed nature of Norwegian Sea IRD and potential endmembers

The diverse and strongly mixed character of every sample demonstrates that the surface currents and iceberg sources changed during the time span represented by each sample. Figure 5.2 illustrates the data distribution and lithologic diversity of the Norwegian Sea IRD. The four most abundant component groups are plotted as two ratios: the ratio of crystalline to sedimentary rocks on the y-axis, and the ratio of quartz to feldspar on the x-axis. The use of ratios allows it to plot four components in one diagram. Clearly the greatest density of data points is in the lower left of the plot, with a group of samples spreading away to the right on the x-axis. These samples are very rich in clastic sedimentary rocks. The sample farthest to the right (23258, 659–661 cm core depth, ca. 14.7 kyr) is from the continental slope of the northwestern Barents Shelf. It contains 97% ice rafted clastic sedimentary rocks from the bottom of the Barents Sea. These clastics were dropped close to the former margin of the Barents Ice Sheet as it disintegrated (Bischof, 1994; Bischof et al., 1997; Weinelt et al., 1991). The Barents Ice Sheet decay will be described in more detail later in this book.

The other extreme is a single data point in the upper left of Fig. 5.2. The data point belongs to a sample from 545–565 cm depth in kastenlot core 23074, located close to the western continental margin of Norway. The sample contains 45.7% crystalline rocks (mostly gneisses), 37.1% quartz, 15.7% feldspar and 9.6% dark green basalts. Quartzites are only 1.7%, clastics 0.7%, and phyllites, chalk and flint are missing. This composition places the origin of this sample firmly to the Archean basement in southwestern Norway, to the south of approximately 62° North. The lack of chalk and flint shows that no rocks from the southern Baltic region were mixed into this IRD, and the lack of phyllites and mica schists excludes central western Norway north of 64°N. Therefore, this particular sample serves as a fingerprint of the Archean basement rocks from southwestern Norway.

But as Fig. 5.2 shows, the samples are not simple mixtures of two or three endmembers. If we concentrate on the area of maximum data point density between ratios of 0–10 (Fig. 5.5a), we find the top half and lower right of the diagram sparsely 'populated'. The true maximum concentration (82% of all samples) is in the area between 0 and 5 for either ratio. Thus, while the extreme outlier samples in Fig. 5.2 represent nearly undiluted, pure source material from distinct endmembers – the northwestern Barents Sea and the southwestern Norwegian Archean formations – the vast majority of the samples is concentrated in a much narrower range of ratios, and more thoroughly mixed with material from many different sources.

In the most trivial case, at least four endmembers are necessary to produce the data points in Fig. 5.5a. These endmembers are the four corners of the diagram. One of these endmembers must have 10-times more crystalline than sedimentary rocks and more feldspar than quartz (top left). This is quite realistic, as any Archean granite and gneiss complex would have abundant crystalline rocks and feldspar. The next one, in the lower left of the diagram, is more problematic, as it must have more sedimentary than crystalline rocks, but also more feldspar than quartz. But the

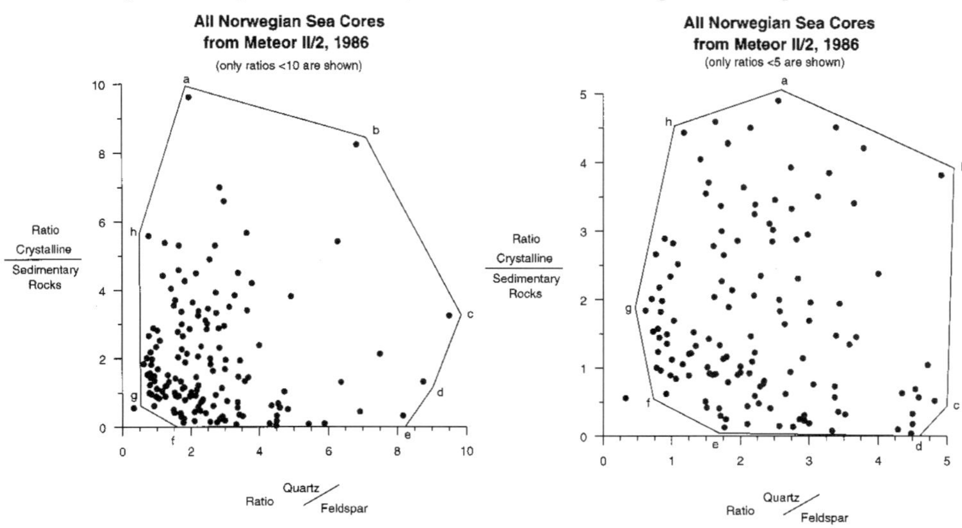

Fig. 5.5. (a) (b)

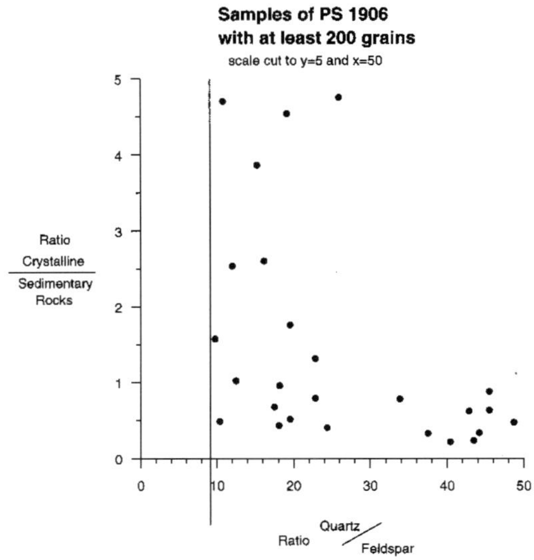

(c)

Fig. 5.5. Composition of >500 μm IRD in the Norwegian Sea as two ratios, the ratio of crystalline to sedimentary rocks (vertical axis), and quartz to feldspar (horizontal axis).
(a) enlarged segment with ratios between 0–10. The data point distribution is relatively loose outside the 0–5 square, but becomes progressively denser towards the lower left corner. Ninety-three percent of the samples are within the octogon (a–h). The corners of the polygon denote possible endmembers of which the IRD assemblages are mixtures; (b) distribution of data points in the range of values from 0 to 5 for each ratio. Eighty-two percent of all samples are within this range, and the new, tighter polygon shows the most realistic endmembers; (c) to illustrate the differences between IRD in the Greenland and Norwegian seas, here the same ratios as in a and b are shown for a core from the East Greenland continental margin (PS 1906). This diagram shows the composition of IRD derived from coastal areas of East Greenland. In addition to containing rocks that have a profoundly different mineralogy (pistaccio-green epidote gneisses, dark green clinopyroxene grains) than those found in Norwegian Sea IRD, this IRD is very different because it has, invariably, at least 10 times more quartz than feldspar grains. But in the Norwegian Sea, the ratios for quartz to feldspar are mostly less than 10. The fact that here the >250 μm fraction was used changes the ratio between rock fragments and single mineral grains, but not the ratios between quartz and feldspar, or between sedimentary and crystalline rocks. Thus, the data are directly comparable and show that the Greenland Sea IRD is different from IRD in the Norwegian Sea.

breakdown of the mostly quartzose clastic sedimentary rocks that make up most of the sedimentary rock IRD would produce clastic sedimentary rock fragments and quartz grains, but almost no feldspar. Only arkoses or meta-arkoses would produce both clastic rock fragments and feldspar grains, but such rocks are very rare among Norwegian Sea IRD.

In fact, all of the data points closest to the x–y origin are offset by a small margin. The data point close to the 'g' should be disregarded because it is based on a Holocene sample with only 27 grains. The next closest point has a ratio of 1 for quartz to feldspar and 0.75 for crystalline to sedimentary rocks. It is thus more reasonable to think of one of these samples as possible endmembers.

The other two endmembers in the lower and upper right are easier to imagine: lower right would be a clastic sedimentary and quartz terrain such as the northwestern Barents Sea, and top right would be high in crystalline rocks with a high quartz to feldspar ratio such as a quartzite terrain like the Caledonian nappes of western Norway.

Even though, as we have just seen, a four-endmember case is possible, it is probably too simplistic to explain the composition of Norwegian Sea IRD. If we keep in mind the diversified bedrock geology around the Norwegian Sea, with all imaginable kinds of crystalline rocks in Norway and the northwestern British isles, mixed sedimentary/crystalline terrains in Svalbard and East Greenland, and even more sedimentary and volcanic rocks on the shelves, and compare this with the extremely high diversity of the Norwegian Sea IRD, a more complex multiple-endmember case must be considered. The data-point cloud of Fig. 5.5a is located within an eight-sided polygon, marked clockwise from a (top left) to h. Ninety-three percent of all samples are located within this polygon. The corners of this irregular octogon are actual samples.

Let us briefly discuss what these endmembers could be. Point 'a' in the top left is a sample[12] that contains 9.6 times more crystalline than sedimentary rocks and twice as many quartz than feldspar grains. This means that (a) the source terrain(s) is composed almost exclusively (perhaps to more than 90%) of crystalline rocks, and (b) these crystalline rocks have a relatively high feldspar content.[13] The granites and gneisses from southwestern Norway could produce these ratios, along with some quartz from the decay of coarse clastics or quartzose meta-sediments.

Point 'b' has a similarly high crystalline to sedimentary rock ratio, but a much larger quartz to feldspar ratio of 6.8. Again, at least 80% of the source areas must be composed of crystalline rocks, but with a higher quartz content than those at point 'a'. This points to a source in the Caledonian nappes of central western Norway or to the Hekla Hoek rocks of Svalbard. Considering that the sample (23269, 0–5cm depth; Fig. 5.1) is from the northern Norwegian Sea and that it has an age of close to the present, the source area must be glaciated at the present, or be located near ice-covered waters. This eliminates most of Norway and leaves us with northern and western Svalbard as the most likely source.

Point 'c' is located at the far lower right margin of the diagram and has a considerably lower crystalline to sedimentary rock ratio. The ratios translate into approximately 75% crystalline and 25% sedimentary rock fragments, and 90% of the mineral grains are quartz. Here, the quartz dominance is increasingly the product of coarse clastic rocks. The sample is 23070, 1–6cm depth (Fig. 5.1), again a Late Holocene sample with a possible age of 3000–4000 years bp. The ice that carried these IRD fragments to the southern Norwegian Sea must have originated from several places in Svalbard, but still with a preference to the northern and western side, which are glaciated and iceberg productive (Dowdeswell and Dowdeswell, 1989).

The next two points, 'd' and 'e', differ from 'c' in that the sedimentary rocks are becoming abundant at the expense of crystalline lithologies. The corresponding samples (23071, 0–8 cm and 23067, 1–6 cm) are again of Late Holocene age and thus contain debris from a similar source as 'c'. But the increasing amounts of sedimentary rocks indicate that the southern part of the Svalbard archipelago (Edgeøya and Barentsøya with their clastic sedimentary rocks) became a more important source of IRD than the northern and western parts.

Point 'f' at the lower left belongs to sample 23059, 310–330 cm. This sample is from the earliest part of glacial isotope stage 6, about 180 ka. More than 88% of the rock fragments are clastic sedimentary rocks, and the remainder are quartzites, meta-argillites, and some gneiss and granite fragments. Clearly, the source for this type of composition must have been a clastic sedimentary terrain with only minor proportions of crystalline material. The clastic sedimentary rocks are more than 50% grey, lithologically uniform siltstones, followed by lesser quantities of white, fine-grained sandstones, black shales and brown siltstones. These rocks are petrographically almost identical to those that came during the late stage 2 from the western Barents shelf (Bischof, 1994; Bischof et al., 1997). But sample 23059, 310–330 cm, has two more characteristics that help define the source: it did not contain a single fragment of coal, nor any volcanic rock fragments, even though both components are common in small quantities in most Norwegian Sea IRD samples. The absence of volcanic rocks eliminates Franz Josef Land as a source, and the lack of coal makes it unlikely, if not eliminates, the possibility of a derivation from Siberia, east of the Taymyr Peninsula. A Kara Sea derivation is also unlikely, because sediments from there contain mostly (70–>90%) quartz grains of more than 250 µm in diameter (Bischof and Darby, 1997). Therefore, the host rocks of these quartz assemblages are most likely coarser than the silt and claystones found in our Norwegian Sea IRD sample. Point 'f' thus represents almost pure Barents shelf debris, and indeed documents an earlier than Weichselian incarnation of the Barents Ice Sheet. This interpretation is further corroberated by the results of cluster analyses, presented in a later chapter.

Point 'h' is an IRD assemblage with a ratio of 5.6 for crystalline to sedimentary rocks and 0.8 for quartz to feldspar. These ratios translate into a source area with 85% crystalline and circa 15% sedimentary rocks, with the crystalline rocks being feldspar rich granites and gneisses. In fact, a low ratio of quartz to feldspar is exactly what one should expect from the disintegration of the light coloured plutonic and metamorphic rocks in Norwegian Sea IRD (see [13]). Therefore, western and southern Norway are the best sources for the granites and gneisses, whereas the sedimentary rocks are from other places. The sample (23062, 31–42 cm) is from the central Norwegian Sea and has an age of close to the last glacial maximum, 25,200–18,600 years before the present.[14] This was the only time during the Weichselian Glaciation when the Fennoscandian Ice Sheet reached the outer shelf, thus overriding sedimentary bedrock (Mangerud, 1991a, b). Therefore, this sample could be an example of crystalline material from land areas and the inner shelf of southwestern

Norway, and sedimentary rocks from the outer shelf. It also sets limits to the amount (percentages) of sedimentary rocks that can be expected from glacial erosion of the outer shelf of southwestern Norway, approximately 6–7%. But the numbers may be different for other parts of the Norwegian shelf where it is wider.

Thus far, the data distribution is in agreement with a derivation from at least three major endmembers: (1) mainland Norway and, at the last glacial maximum, the Norwegian shelf, (2) the Barents Sea (Barents Ice Sheet), and (3) various places in the Svalbard archipelago. The samples are mixtures, composed mainly from these major sources. Additional debris came from secondary sources such as the southern Baltic region (chalk and flint), and from North Greenland (black shales with Lower Paleozoic acritarchs). Of those minor sources, the Baltic region has consistently delivered to all locations in the Norwegian Sea, whereas North Greenland's influence is only faintly recognisable in the western Norwegian Sea. Contributions from East Greenland to the south of 80° North are not recognised in IRD assemblages of the Norwegian Sea. This is because a core from the East Greenland continental margin contains IRD with a ratio of at least 10 for quartz to feldspar (Fig. 5.5c), but in the Norwegian Sea this ratio is mostly less than 5.

But let us now discuss whether more detailed endmember determinations are possible. The upper left side of Fig. 5.5a symbolises the Norwegian-derived proportion, the right side material from Svalbard, and the bottom axis debris from the Barents Sea. The data point distribution within the polygon is very irregular. By far the majority of samples (82%) are within the lower left square of ratios from 0–5, and only 18 samples are in the periphery. In order to see which endmembers are responsible for most of the samples, we concentrate on the 0–5 area (Fig. 5.5b). Here, the data points are almost evenly distributed, which clearly indicates the complexity of the composition of the Norwegian Sea IRD. The corners of the bounding polygon a–h represent possible endmembers. Here, at least eight endmembers are necessary to produce the data diversity.

Application of this graphical concept of endmember identification to other IRD lithologies reveals that practically any type of IRD lithology is composed from many different sources. Figure 5.6a shows the composition of clastic sedimentary rocks expressed as two ratios. Again, the data points are densely clustered in the lower left of the diagram with a few extreme outliers. Figure 5.6b is an enlargement of this area. In this case, the bounding polygon suggests at least eight end members, a–h. The starting point, a, has ca. 14 times more grey than black clastics, and twice as many orange, red and yellow than brown clastics. On the other hand, f and g have more black shales than grey clastics, and contain more brown clastics. Data points on the left axis are samples devoid of brown clastics, which effectively rules out the Barents shelf as a source for these samples, because brown clastics (primarily siltstones) are always abundant there. But the right side, particularly the upper right, points c and d, could represent debris from the Barents shelf with grey clastics being 9–13 times more frequent than black shales, and brown clastics being 2–5 times more abundant than orange, red and yellow ones.

For quartzites, the data distribution looks very similar at first sight (Fig. 5.7). Again there is a maximum concentration in the lower left with outliers in the top left

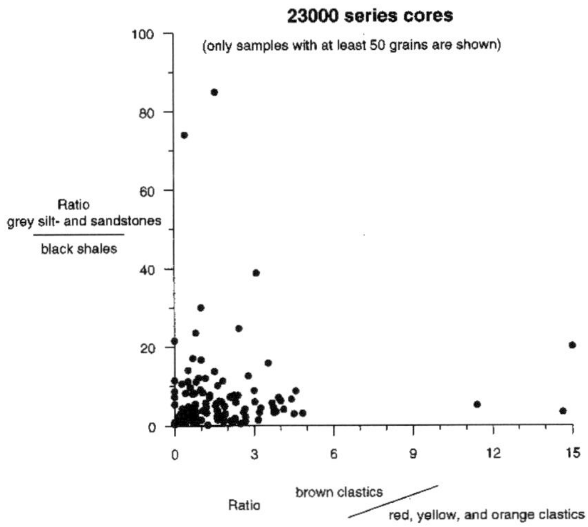

Fig. 5.6. Ratios of the most frequently occurring types of clastic sedimentary rocks, by colour and grain size. The graph shows a tight cluster in the lower left corner and two extreme outliers. At the top left are samples in which grey silt- and sandstones are 70–90 times more frequent than black shales, and near the bottom right samples contain up to 15 times more red, orange and yellow than brown clastics. But these extremes are rare, and most samples have smaller ratios.

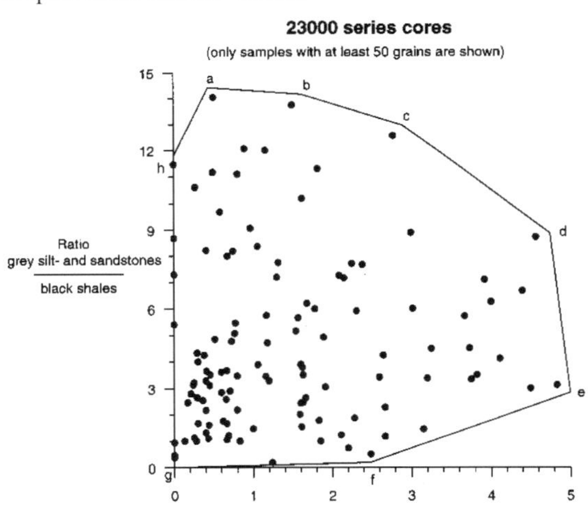

(b) The lower left of Fig. 8d enlarged. The points are almost evenly scattered across the diagram. Again, the points are surrounded by an eight-sided polygon which represents a relatively simple eight endmember case.

and bottom right (Fig. 5.7a). The sample with the highest ratio (almost 12) of green quartzites to red, orange, yellow and brown ones is from core 23068 and the middle part of stage 6, the next largest glaciation prior to the Weichselian. The sample is dominated by fine-grained clastics and contains 20% coal fragments with an origin somewhere in Siberia (Bischof et al., 1990). It is therefore possible that these green

(a)

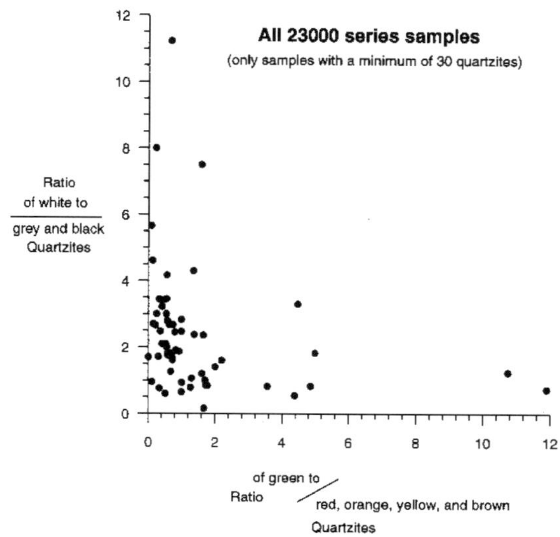

(b)

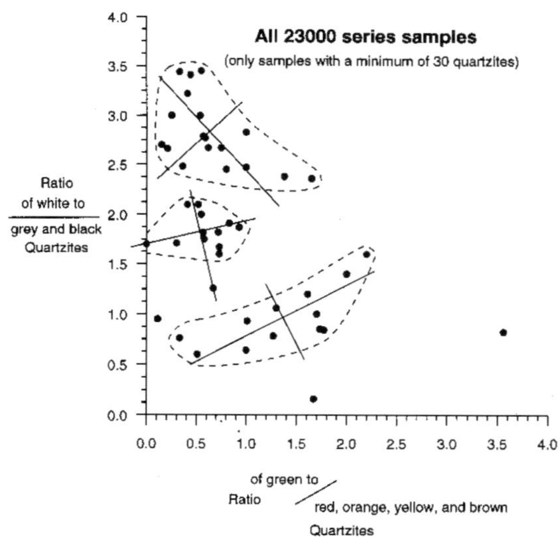

Fig. 5.7. Ratios between quartzites of different colour. Similar to the ratios of clastic rocks, there are two extreme endmembers at the top left and bottom right. At the top left are samples with abundant white – clear, red, orange, yellow and brown quartzites, and at the far right are samples with high amounts of green, grey and black quartzites. (b) Enlargement of the lower left of Fig. 5.7a. Unlike the aforementioned ratios, the quartzite ratios do not fill the plotted area evenly, but form three distinct groups. Each of these groups represents a quartzite terrain of different composition. The natural variability within each of those source terrains is illustrated by the orthogonal coordinate systems drawn through them.

quartzites typify quartzite outcrops close to the origin of the coal in Siberia, possibly the Taymyr Peninsula (Churkin et al., 1981). The data point at the top left belongs to a sample of Middle Weichselian age from the eastern Norwegian Sea. It is dominated by white quartzites and contains green, orange, red and yellow quartzites in even amounts. This type of composition possibly characterises IRD from the Norwegian Caledonides.

But if the lower left is enlarged (Fig. 5.7b), a data structure emerges that considerably deviates from those seen before. Here, only the triangular left side of the square is occupied, and not evenly, but with data points concentrated in three distinct groups. Within each of these groups, the data points are located in an area formed by two orthogonal axes. These axes describe the variance within each group, and are nothing more than a graphical, and rather primitive, approach towards factor analysis. It is a dimension reduction technique designed to find a limited number of factors (or endmembers) in a multivariate data set. We have identified some of the major and minor source areas already and will explore later whether more differentiation is possible with factor and cluster analysis.

Feldspar grains are an important component of Norwegian Sea IRD and set the Norwegian Sea apart from the Arctic Ocean and the Greenland Sea where feldspar is generally unimportant. Figure 5.8a shows the colour distribution in a ternary diagram. The feldspars are mostly white and clear, and all data points are concentrated in the lower right part of the figure. This points to a rather uniform and consistant source of the feldspars. Figure 5.8b shows the same properties of light coloured intrusive rocks, granites, tonalites, diorites etc. They have a similar colour distribution as the feldspar grains, but the area occupied by the data points is larger and suggests a greater diversity of the sources of the plutonic rocks. Also, the centre of the data points has shifted to the left, towards more red and orange intrusives. The point clouds of the feldspars and intrusive rocks only partly overlap. Therefore the feldspars and the light coloured intrusive rocks are to some extent from the same origin, and to some extent from different source terrains. The data point distribution suggests in both cases five or six possible endmembers.

We are now trying to explore whether there are specific sub-regions within the thus far determined IRD sources that have dominated the iceberg drift periodically. Delineating such sub-sections, of continental Norway for example, is important because it will give a clue about how the surface circulation and ice drift changed through the time spans represented by our samples. It can provide an image of the variability of the Pleistocene circulation. For example, demonstrating that northern Norway was at times a very influential IRD source for the eastern Norwegian Sea would amount to proof of southward currents, opposite to the main direction of the surface currents today. But to distinguish between northern Norwegian crystalline rocks and crystalline rocks from other places in Norway is extremely difficult. Even though the general bedrock geology is quite well known, the level of detail of the petrographic descriptions is usually not high enough to allow a comparison with the >500 μm IRD studied here. What is needed is information that would allow us to imagine what the debris from a particular rock formation would look like after it has been overridden by glacial ice, crushed and broken down to fragments of the size in

(a)

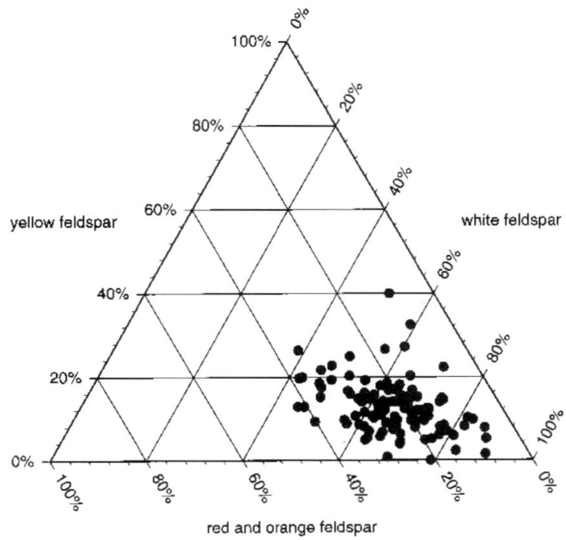

(b)

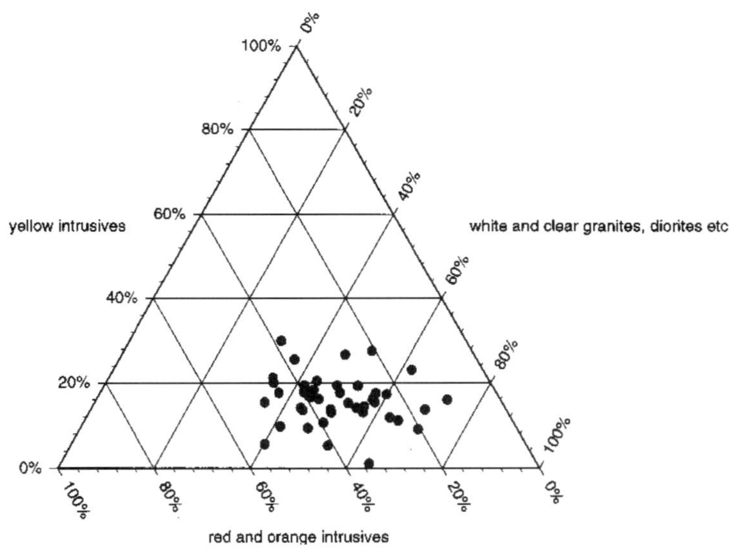

Fig. 5.8. Feldspar grains by colour. All data points are concentrated in the lower right side of the ternary diagram. White and clear feldspar (mostly plagioclase, but also microcline) is dominating over red, orange, and yellow varieties. The samples are mixed from at least three or as many as five different source terrains.
(b) Same ternary plot for light coloured intrusive rocks such as granites, tonalites, syenites etc.. The colour distribution is similar to the one of feldspar, but the field is wider and the graphic centre offset to the left. This points to a partly syngenetic and in part different origin of the feldspar grains and granite fragments.

the Norwegian Sea IRD. In the classification scheme used by this author, grey granites would fall apart into fragments of white granite, diorite, quartz and white feldspar, and occupy the lower right corner in the ternary diagrams of Fig. 5.8.

But it is possible to narrow down the potential sources, even in a semi-quantitative way. Metapelites provide an important clue: their occurrence in Norway is confined to the Caledonian nappes which consist of various kinds of metasediments of variable metamorphic grade. Between approximately 64° and 69°30 North, the dominant surface-exposed rock types are metapelites, phyllites and mica schists (Sigmond et al., 1984). Judging from their geologic map, approximately 60–70% of the surface area is composed of these metapelites, which also have been intruded by granite plugs. Therefore, after this area has been overridden by erosive glaciers, one should expect circa 60–70% of the resulting glacially-eroded debris to be mica schists and phyllites, the remainder being granite fragments, quartz and feldspar. And because the >500 μm fraction usually consists of one half rock fragments and one half minerals (quartz and feldspar), the final composition of iceberg rafted debris, if it had been entirely derived from this section of western Norway, should be 30–35% metapelites, 10–15% granites, and 50% quartz and feldspar. But the highest percentage of such metapelites in Norwegian Sea IRD is only 11.1% from a box grab from the shelf of northern Norway. This sample is located closest (only 200 km offshore) to the presumed place of origin and shows that central-western Norway between 64° and 69°30 North is indeed the source of the phyllites and mica schists. But the decrease of the metapelite concentration from a presumed 30–35% to about 10% shows that even within a relatively short distance of 200 km, the original concentration of debris from central western Norway was diluted to circa one-third owing to advection of ice rafted debris from other sources. With increasing distances from coastal Norway, the metapelite concentrations drop further and the contribution from this Norwegian source becomes very small. But on the other hand, contributions from central western Norway are always recognisable because of the continued presence of the metapelites, and because of their very uniform appearance. They consist of only two types of rock: grey, platy, low metamorphic phyllites, and silver mica schists of higher metamorphic grade. The metapelites lack the colour variety of those found to the north of Ellesmere Island in the Arctic Ocean and in the northern Greenland Sea.

The other samples with the next highest percentages of phyllites and mica schists (4.2–9.1%) are either middle Holocene samples from the eastern Norwegian Sea, Late Pleistocene (18–23 ka) samples from the western Norwegian Sea, or samples of uncertain stratigraphic position from the central Lofoten Basin (Table 5.2). In particular the presence of metapelites in sediments of middle Holocene age is interesting because this is generally regarded as a relatively warm climatic period. These metapelites could have originated only from two possible sources: (a) coastal northern Norway north of 65 degrees, or (b) from somewhere in the Arctic Ocean. Whereas a derivation from the Arctic, from Svalbard or the Taymyr Peninsula for example, is quite plausible owing to the continued presence of an Arctic sea ice cover, an extraction from northern Norway would require at least winter ice formation or even glaciers reaching sea level. But whichever of the latter scenarios is

Table 5.2. Meta-argillites listed in decreasing order

Poseidon cores	Top	Bottom	Meta-argillites in percent of all >500 μ IRD	Meta-argillites in percent of all rock fragments
16315	0	50	11.08	12.53
16313	0	50	4.73	9.11
16061	0	4	3.79	8.06
16072	12	25	3.97	6.95
16061	8	12	2.92	5.98
16061	4	8	2.82	5.71
16072	4	12	3.60	5.54
16061	12	17	2.50	5.29
16327	0	50	4.92	5.02
16061	20	23	2.15	4.39
16061	26	29	1.95	4.18
16061	29	32	1.86	3.82
16072	0	4	2.44	3.79
16061	17	20	1.71	3.74
16052	5	8	1.71	3.61
16069	2	5	1.42	3.44
16061	32	34	1.61	3.35
16069	35	39	1.35	3.23
16069	0	2	1.60	3.08
16069	33	35	1.30	2.97
16365–1	15	16.5	1.62	2.83
16043	10	14	1.11	2.69
16052	2	5	1.67	2.64
16043	2	5	1.13	2.42
16061	23	26	1.13	2.41
16052	0	2	1.23	2.40
16052	8	14	1.38	2.34
16043	0	2	1.15	2.30
16043	23	26	1.06	2.28
16069	23	26	1.38	2.25
16069	16	20	1.21	2.06
16052	17	21	1.55	2.06
16365–1	24.5	26	1.06	2.04
16365–1	20	21.5	1.09	1.98
16366–2	34.5	36	0.80	1.91
16069	29	33	1.00	1.83
16069	10	13	1.24	1.83
16365–1	38	39.5	1.01	1.80
16061	34	36	1.01	1.80
16043	20	23	0.86	1.76
16069	5	8	0.99	1.75
16052	35	41	0.75	1.70
16072	30	35	1.25	1.67

Table 5.2 *cont.*

Poseidon cores	Top	Bottom	Meta-argillites in percent of all >500 μ IRD	Meta-argillites in percent of all rock fragments
16073	30	40	1.26	1.62
16052	29	32	0.76	1.61
16366–2	15	16.5	0.96	1.58
16043	7	9	0.70	1.46
16366–2	28.5	30	0.53	1.44
16069	26	29	0.93	1.43
16043	5	7	0.57	1.35
16366–2	39	40.5	0.58	1.34
16052	14	17	0.92	1.33
16366–2	44	46	0.68	1.30
16069	13	16	0.79	1.27
16069	20	23	0.88	1.19
16043	14	17	0.74	1.17
16366–2	19.5	21	0.73	1.13
16365–1	43	45	0.80	1.08
16365–1	29	30.5	0.63	1.06
16052	26	29	0.76	1.01
16073	0	10	0.70	0.99
16052	21	26	0.75	0.91
16043	28	30	0.40	0.80
16366–2	24	25.5	0.68	0.75
16073	20	30	0.47	0.72
16043	30	35	0.37	0.71
16073	10	20	0.42	0.57
16073	40	47	0.42	0.53
16072	25	30	0.36	0.52
16043	26	28	0.24	0.50
16365–1	33.5	35	0.34	0.48
16052	32	35	0.22	0.48
16043	17	20	0.23	0.46
16069	8	10	0.20	0.33
16043	9	10	0.09	0.18
16043	0	2	0.00	0.00
16315	0	50	0.00	0.00

Concentration of grey phyllites and silver mica schists in glaciomarine IRD of the Norwegian Greenland Sea. Shown here are the 16000 series cores that were recovered by RV *Poseidon* in 1987 and previous years. The table shows, from left to right, core number, depth of the sample at the top and bottom, the percentage of metapelites of all >500 μm IRD, and of all rock fragments. Samples 16313 and 16315 with highest metapelite percentages are dredged samples from the northern Norwegian shelf north of 67° N. They prove that the ultimate origin of the metapelites is from this part of Norway, roughly between 65 and 68 degrees north.

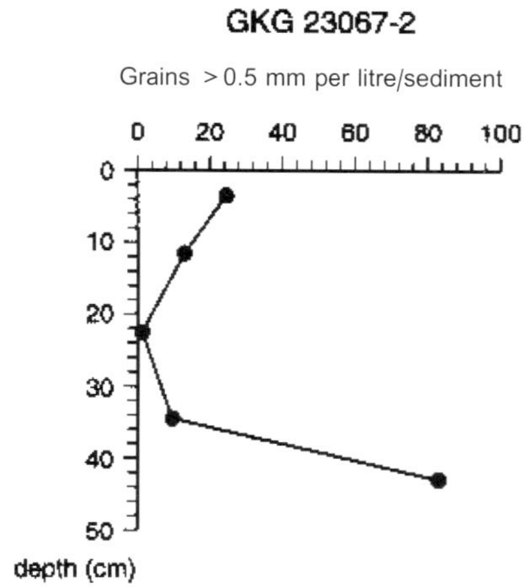

Fig. 5.9. Concentrations of grains >0.5 mm per one litre sediment in core 23067–2.

preferred, it would amount to colder conditions than at the present, and consequently, sediments deposited at this time should contain elevated concentrations of ice rafted coarse grains. But the opposite is the case: middle Holocene sediments generally contain the lowest grain densities of the Holocene (Fig. 5.9). Therefore, the idea of a northern Norwegian provenance of the mid-Holocene ice rafted metapelites must be rejected. They were derived from either northwestern Svalbard or from the Taymyr Peninsula.

This allows us to conclude that (1) during the Late Weichselian glaciation, before the Barents ice sheet disintegrated, icebergs from central western Norway drifted in a wide arc to the west and northwest, reaching sites in the western Norwegian Sea; (2) during the middle Holocene, ice drifted from northwestern Svalbard and/or the Taymyr Peninsula southwards to the eastern Norwegian Sea. The type of ice could have been Arctic pack ice or icebergs.

Finding ice rafted debris from a northern Norwegian source in Late Weichselian sediments is not a big surprise because Norway was completely glaciated at that time. But finding IRD from this same source in sediments that are undeniably of middle Holocene age (time range: approximately 6000–4000 years bp) *is* a big surprise, because this time period is commonly considered one of the warmest intervals of the Holocene.

In the view of this author, these ice rafting episodes most likely represent brief cold snaps that occasionally pierced longer-lasting intervals of continued warmth. They represent a climatic stade in which the North Atlantic Drift–Norwegian Current system was weaker than normal, perhaps in response to or as a cause of a reduction of the formation of North Atlantic deep water. As a consequence of the weakened northward-moving waters, cold Arctic surface waters with sea ice and possibly also icebergs could penetrate south and cover vast parts of the Norwegian Sea. The aereal extent of winter ice in the Norwegian Greenland seas thus was highly variable during the Holocene. But more on that subject in Section 5.5.

5.3.4 Cluster and factor analysis

To improve and refine the source area determinations made above, cluster and factor analyses were performed with 252 samples and 45 variables (=IRD lithologies). Both statistical procedures are designed to find specific sub-sets of samples characterised by high mutual similarity (cluster analysis), or to determine whether the composition of a very complex set of data is composed of a limited number of contributing factors, or source areas (factor analysis). Both techniques mathematically begin by converting the data, in our case the 45 lithologic variables, into vectors in a 45-dimensional vector space. In a simple three-dimensional case, the vectors are straight lines with the same origin, oriented at different angles in a sphere. The program then searches for samples that are located close together in this vector space ('bundles' of vectors close together, so to speak) and combines them into groups of very similar composition, clusters. For more on the mathematical background, the reader is referred to Davis (1986).

5.3.4.1 Cluster analysis

The procedure fastclus (SAS Institute, 1989), also referred to as the 'nearest centroid sorting' method (Anderberg 1973), was chosen. Procedure fastclus is an iterative process that initially selects some observations (samples) as cluster seeds around which other samples are clustered, calculating the distances between samples on the basis of Euclidian distances with all 45 input variables. The criterion for assigning a sample to its respective cluster is that the distances between observations in the same cluster are smaller than all distances between observations in different clusters. The program then calculates cluster means as new seeds and repeats the clustering until no further changes occur. This method ensures that clusters of similar observations are always found, even if the differences between adjacent clusters are small. The latter aspect proved to be crucial in distinguishing the clastic sedimentary rock suites of the Barents shelf from other areas of similar bedrock composition in circum-Arctic regions.

The objective of cluster analysis with the Norwegian Sea IRD data was to find groups of samples whose composition is sufficiently similar that they must have originated from the same general source. The first example is cluster 22 (the original cluster numbers as produced by the SAS procedure are used here), randomly selected out of a total of 34 clusters with at least two members (Table 5.3). The samples have a middle to late Weichselian age, are located in the central and eastern Norwegian Sea, and share a highly diverse IRD assemblage with dominating gneisses. Five of the samples are older than the Last Glacial Maximum and range in age from 66.5 to 25 ka, while one sample from core 23064, central Norwegian Sea, has an age of 13.6 to 12 ka.

The IRD composition points to a source in southwestern Norway with mostly metamorphic rocks and some debris from the southern Baltic area. The main source area must have been composed of primarily light coloured, fine–medium crystalline metamorphic and igneous rocks and quartzites, whereas the proportion of phyllites and mafic rocks was relatively small. This composition points to one or several of the Precambrian complexes or the Caledonian orogenic belt of western and southern

Table 5.3. Cluster 22

Observation	222	224	238	239	240	60
sample# (SAS)	NGS272	NGS274	NGS288	NGS289	NGS290	NGS76
core #	23068	23068	23071	23071	23071	23064
top (cm)	96	170	240	370	480	12
bottom (cm)	119	180	260	385	500	20
age top (years bp)	25000	38000	25000	43600	59000	12000
age bottom (years bp)	27000	40000	27000	45700	66500	13600
thickness (cm)	23	10	20	15	20	8
time of depo.	2000	2000	2000	2100	7500	1600
cluster	22	22	22	22	22	22
distance from cluster seed	0	11.6	14.4	12.3	12.8	10.3
white sandstones	1.9	1.06	0.31	1.56	1.68	2.28
grey siltstones	5.12	5.55	4.92	6.26	8.16	7.11
grey sandstones	0	0	0	0	0	0
black shales	0	2.11	0.73	0.7	1.94	4.3
green siltstones	0.19	0.13	0.21	0.17	0.39	0.13
brown siltstones	2.85	0.66	1.57	0.61	1.94	2.15
red siltstones	4.17	2.77	0.52	1.22	2.52	2.68
clay-ironstones	0.19	0	0.21	0	0	0
other clastics	0	0	0	0	0	0
limestones and dolomites	2.66	0.4	2.62	0.43	1.23	0
chalk	0	0	1.78	0.09	0.32	0
chert	0.57	0	0.1	0.17	0.39	0.13
flint	0.38	0	0.94	0.09	0.06	0
white granites	4.37	0.13	1.05	2.34	0.32	5.37
red granites	1.52	0	0.41	1.21	0.39	2.14
yellow granites	0.76	0	0.1	0	0.19	1.74
white/red granites	0.19	0	0.31	0.34	0	0.67
other granites	0	0	0	0.09	0	0.67
gabbro	2.28	0	1.25	0.52	0.51	2.01
pyroxenites	0	0	0	0.17	0.13	0.27
phyllites and mica schists	2.28	1.05	0.94	2.26	1.87	3.49
serpentinites	0	0	0	0	0	0
white quartzites	4.55	2.51	10.56	7.3	4.14	4.3
grey quartzites	1.71	2.25	0.94	2.17	1.75	4.03
green quartzites	0.76	0.26	1.57	0.7	1.68	2.28
red quartzites	2.28	1.06	1.26	0.52	1.29	0.81
other quartzites	1.33	1.72	1.04	1.04	1.55	1.61
light gneiss	23.14	28.53	17.48	21.63	19.22	19.07
mafic gneiss	14.61	15.46	11.3	11.03	9.39	8.99
volcanic glass	0	0	0	0.09	0.19	0
dark volcanics	1.52	1.06	1.25	0.26	3.69	0.54
red volcanics	0.19	0.26	0	0.17	0.64	0.26
pumice	0	0	0	0	0	0
quartz	15.94	21.14	21.98	22.85	17.79	15.98

Table 5.3 *cont.*

Observation	222	224	238	239	240	60
white feldspar	2.47	7.13	9.1	8.52	10.1	5.37
red feldspar	1.14	2.91	3.45	4.26	4.52	0.54
yellow feldspar	0.76	1.58	2.09	1.13	1.74	0.66
light green feldspar	0	0	0	0	0	0
garnet	0	0.13	0	0.09	0.13	0
mafic minerals	0	0	0	0	0.06	0.4
olivine	0.19	0	0	0	0.06	0
black opaques	0	0	0	0	0	0
coal	0	0.13	0	0	0	0
biotite	0	0	0	0	0	0
muscovite	0	0	0	0	0	0

This table presents the IRD composition of all six samples that were grouped into Cluster 22, one of 34 clusters in total. The six samples are from different stratigraphic levels of three cores from the eastern and central Norwegian Sea. The top portion of the table provides basic information about the samples (depth, age, etc.), and the lower part gives the IRD composition (grains > 500 μm) in grain percent. All samples contain a wide array of different lithologies in similar proportions with crystalline components, particularly multicoloured gneisses, dominating. Southwestern Norway is the dominant source, along with minor contributions from the southern Baltic (chalk and flint). Sample 23068, 170–180 cm, deviates from the others in that it is practically devoid of plutonic, intrusive rocks (shaded area), and the crystallines are almost entirely metamorphic rocks. Chalk and flint are absent (shaded cells). Therefore, this sample is nearly pure debris from a joint Precambrian gneiss–quartzite–metapelite terrain: southwestern Norway without, or with very little, debris from the Baltic.

Norway or the northwestern British Isles. The composition excludes, however, areas which are composed dominantly of phyllites and mica schists and/or mafic rocks, such as the area north of the Lofoten Islands where mafic rocks dominate (Sigmond et al., 1984; Oftedahl 1980), the area between the Trondheim Fjord and the Lofoten Islands (phyllites and mica schists), the Bergen Arch (mafic rocks), and the outer Hebrides (mafics, basalts). The most likely sources are therefore the Precambrian gneisses and Caledonian nappes of central western Norway south of the Trondheim Fjord. The samples contain small but significant proportions of limestones, dolomites, chalk and flint as indicators of partial influx from the southeastern Norwegian Sea. This extends the sources for samples of cluster 22 to southern Sweden and the Baltic region.

The cluster reveals that this source contributed throughout most of the mid-to pre-late Weichselian, and that an ice drift direction from the southeastern Norwegian Sea to the north was operative for more than 40,000 years. However, it must be reiterated that the three samples from kastenlot core 23071, which are in consecutive order, were taken from the coarsest sections. They therefore represent the composition of IRD during times of maximum iceberg 'production' from the southwestern Fennoscandian Ice Sheet.

Table 5.4 shows a cluster that contains one sample with debris from the bottom of the western Barents shelf (23259, 129–130 cm core depth) and five samples from the Norwegian Sea belonging to four different time periods. Three of those time periods are glacial stages: late stage 2 (Late Weichselian resp. Late Wisconsinan Glaciation), middle stage 6, and mid-to-late stage 10. Therefore, the combined Barents/Svalbard Ice Sheet had, in addition to the Late Weichselian edition, at least two earlier incarnations. Each time the icebergs drifted to at least 68° North in the Norwegian Sea. Interestingly, the same ice drift route became reactivated during the latest Holocene (sample 23067, 1–6 cm).

However, the Norwegian Sea samples always contain debris that could not have come from the bottom of the western Barents shelf. These are all rock types in the grey shaded area of Table 5.4: carbonates, chalk, flint, granites, gabbros, quartzites, volcanic rocks and gneisses. These lithologies are missing in the Barents shelf sample but occur in percentages from 7 to 22% in the samples of the Norwegian Sea (Table 5.4, bottom line: 'non-Barents IRD').

There are three possible explanations:

1. The source area(s) that produced the icebergs, or the extent of glaciation, extended beyond the western Barents shelf and included parts of Svalbard where the bedrock consists of crystalline and carbonate rocks.
2. The iceberg drift directions changed during the time periods the samples were deposited (400 to 10,000 years) and allowed icebergs from crystalline and carbonate terrains to drift to the sample sites.
3. Lateral mixing between two surface currents moving in the opposite direction could have caused mixing with lithologies from very different (southern?) origin.

Two of the five samples (23059, 580–590 cm and 23068, 500–520 cm) contain chalk and flint and thus contain advected debris from as far south as the Baltic region. These same samples are also the ones with relatively more non-Barents-derived IRD (11.6 and 14.2%) than the two other Pleistocene samples without any chalk or flint (7.1 and 8.6%). For the two samples with chalk and flint, possibilities 2 and 3 from the above list are thus likely. But a distinction between these two possibilities is not possible, because the samples were 10 and 20 cm thick sections that were homogenised and sampled as one coherent unit. It is thus possible that the IRD composition, and therefore the ice drift directions, changed during the according time interval.

The next important cluster 46 contains 24 samples, of which 19 are from the eastern Norwegian Sea (cores 16043, 16052, 16061, 16069, 23053, 23055, 23073 and 23074). Six samples are from core 23074 from close to the Norwegian continental slope. The cluster, most interestingly, contains also a Late Weichselian till sample from a coastal cliff in northern Germany (from Surendorf near Kiel). This till was bulldozed south by glaciers from central and southern Sweden and contains an eclectic collection of rocks from Sweden and the bottom of the Baltic Sea. The fact that cluster analysis grouped this sample together with 19 samples from the eastern Norwegian Sea which range in age from 50,000 to approximately 10,000 years bp corroberates that icebergs from southern Scandinavia found their way into the

Table 5.4. Cluster 7

Observation	27	177	211	217	230	65
sample# (SAS)	NGS13	NGS227	NGS261	NGS267	NGS28O	NGS87
core #	23055	23259	23059	23062	23068	23067
top (cm)	30	129	580	305	500	1
bottom (cm)	31	130	590	320	520	6
age top (years bp)	16300	13200	340000	155000	163000	205
age bottom (years bp)	16725	13250	350000	165000	166000	1229
thickness (cm)	1	1	10	15	20	5
time of depo.	425	50	10000	10000	3000	1024
cluster	7	7	7	7	7	7
distance from cluster seed	12.18	14.96	10.22	8.12	0.00	16.88
white sandstones	12.07	9.5	5.95	5.51	3.15	6.86
grey siltstones	56.3	56.21	63.68	65.92	61.78	47.06
grey sandstones	0	0	0	0	0	0
blackshales	5.67	11.61	6.27	1.84	2.63	2.29
green siltstones	0.37	0	0.48	1.47	0.39	0.33
brown siltstones	1.65	5.8	2.73	4.17	1.58	1.31
red siltstones	0.18	0	2.57	0.86	0.26	0.65
clay-ironstones	1.65	1.32	0	0.37	0	0.65
other clastics	0	2.37	0.8	0.12	0.39	0.33
limestones and dolomites	0.91	0	0	0.37	1.05	1.31
chalk	0	0	0	0	0.26	0
chert	0	0	0	0.37	0.13	0.33
flint	0	0	0.48	0	0	0
white granites	0.92	0	0.32	0.99	0.26	1.31
red granites	0.91	0	0.16	0.12	1.05	1.63
yellow granites	0.18	0	0.32	0.25	0	0.33
white/red granites	0	0	0	0.12	0	0
other granites	0.18	0	0	0	0	0
gabbro	0.74	0	0.16	0.25	0	0
pyroxenites	0	0	0	0	0	0
phyllites and mica schists	0.55	0.26	0.32	0.37	0.52	0
serpentinites	0	0	0	0	0	0
white quartzites	0.55	0	0.8	1.59	1.05	2.61
grey quartzites	0	0	3.54	0.25	0.92	0.65
green quartzites	0	0	0.8	0.12	0.53	3.59
red quartzites	0.18	0	0	0	0.39	1.31
other quartzites	0	0	1.93	0.24	0.13	0
light gneiss	1.65	0	1.45	0	3.81	5.56
mafic gneiss	0.73	0	1.61	2.08	4.47	2.29
volcanic glass	0	0	0	0	0	0.33
dark volcanics	0.18	0	0	0	0.13	0
red volcanics	0	0	0	0	0	0
pumice	0	0	0	1.84	0	0.65
quartz	11.89	11.87	4.01	8.46	10.64	16.02

Table 5.4. *cont.*

Observation	27	177	211	217	230	65
white feldspar	1.28	0	1.45	1.23	2.63	1.31
red feldspar	0	0.26	0	0.37	0.79	0.65
yellow feldspar	1.09	0	0	0.37	0.13	0
light green feldspar	0	0	0	0	0	0
garnet	0	0.26	0	0.12	0.13	0
mafic minerals	0.18	0	0	0.12	0.13	0
olivine	0	0	0	0	0	0
blackopaques	0	0	0	0	0	0
coal	0	0.53	0.16	0.12	0.66	0.65
biotite	0	0	0	0	0	0
muscovite	0	0	0	0	0	
minimum of non-Barents IRD	7.13	0	11.57	8.59	14.18	21.9

The importance of this cluster lies in the fact that it unites one sample from the Barents shelf slope with five samples from the Norwegian Sea. Sample 23259, 129–130 cm, represents nearly undiluted Barents shelf debris of fine-grained clastic sedimentary rocks (note the almost complete lack of crystalline lithologies in the grey shaded boxes), which reached five different locations in the Norwegian Sea at four different time periods: the latest Holocene, 1300 to 200 years bp; after the last glacial maximum, 16.7 to 16.3 ka; during the middle of stage 6, 166 to 155 ka; and during late stage 10, 350 to 340 ka.

eastern Norwegian Sea frequently during the Weichselian Glaciation, following a northward drift route as shown in Fig. 5.4.

The clusters 13, 18 and 28 are closely related to each other and include 25 samples dominated by clastic sedimentary rocks. The percentage of grey siltstones varies between 27.8 and 53.3%. Cluster 13 contains on average 73.9% clastic sedimentary rocks, while cluster 18 averages 53.2%, and cluster 28 averages 58.5%.

Ten samples of these clusters are from the northern, northwestern and central Vøring-Plateau, eastern Norwegian Sea. Five samples are from the western Norwegian Sea. The central and eastern Norwegian Sea and the western Barents shelf each contributed three samples, and one sample is from the southern Greenland Sea. All samples from the western Norwegian Sea are from core 23059 and older than Weichselian. Their time ranges are 398–367 ka (middle-late stage 11), 257–249 ka (late stage 8), 173–164 ka (early stage 6), 150–141 ka (middle stage 6), and 128–99 ka (stage 6/5 transition – early stage 5). The central Norwegian Sea samples are from 374–347 ka (late stage 11–early stage 10), 183–177 ka (early stage 6), 130–119 ka (stage 6/5 transition–early 5). The samples in the eastern Norwegian Sea are from the time intervals 181–173 ka (early stage 6) and 30.4–26 ka (late stage 3). All Vøring-Plateau samples belong to the time interval from 14.6–8.9 ka.

This is the second group of clusters that combines late Weichselian samples from west of the Barents shelf with samples from the northern-central Vøring-Plateau, and in turn these are linked by their composition to samples from the western-central Norwegian Sea during earlier glacial stages, primarily the stages 11/10, 8 and 6. This is evidence that the western Barents shelf (a) was glaciated at least three times prior to the Weichselian, (b) was the main source of iceberg-rafted debris to the eastern Norwegian Sea during the late Weichselian and (c) was the main source of ice rafted debris to the central and western Norwegian Sea during the older glacial stages 6, 8 and 10.

It is unclear whether the change in the distribution of clastic rock IRD between the older glaciations and the Late Weichselian was a shift of the prevalent iceberg drift direction toward the southwest so that the icebergs missed the Vøring-Plateau, or a westward broadening of the southward currents that carried the icebergs from the Barents shelf to the Vøring-Plateau. The two long kastenlot cores from the southern Vøring-Plateau (23071 and 23074, Fig. 5.1) that could provide the answer to this question do not reach back to stage 6 because of high sedimentation rates.

Samples from the long kastenlot cores were preferably taken from intervals with high quantities of ice rafted debris, whereas adjoining sections with finer grained sediment remained unsampled. The analysed intervals thus represent (relatively short?) periods of high dropstone deposition rates and intensive iceberg rafting, probably surge events, while the dropstone composition in the ambient sediment which represents the 'normal' glacial background sedimentation would have been completely missed. For the same reason, the real time of deposition of the analysed coarse sections could be significantly shorter than the time ranges presented here, because linear interpolation between oxygen isotope stage boundaries was used. In any case, this is a clearly discernable cluster group which unites samples that received their clastic debris from icebergs that originated at the seaward terminus of a grounded ice shield in the Barents Sea. The impact of this ice shield on glacial sedimentation in the Norwegian Sea is now documented for the oxygen isotope stages 11, 10, 8, 6, 3 and 2.

5.3.4.2 Factor analysis

Factor analysis is a statistical method by which a complex data set can be disassembled into a limited number of factors. In case of the Norwegian Greenland Sea IRD samples, the factors represent potential source areas, and their scores (= the projection of every sample on each factor) reveal indirectly the proportion of IRD derived from each source. It is important that the factors are axes in an orthogonal, n-dimensional vector space. Each factor is completely independent from the others: one factor is not a combination of several of the other factors.

Factor analysis produced 13 main factors with eigenvalues of more than one. There were more factors, but factors with lower eigenvalues are insignificant. Table 5.5 shows the loadings of each variable (IRD lithology) on each of the 13 most important factors. From these loadings alone, it is possible to point to the most likely, and in some cases even the only, source they represent.

Let us consider factor 1 first: it has highest loadings (grey shaded cells) for red, yellow and white feldspar, followed by lower values for white, yellow and red granites, and chalk (Table 5.5). This factor represents a source area composed

Table 5.5. Loadings of the variables on the 13 most important factors

	Factor1	Factor2	Factor3	Factor4	Factor5	Factor6	Factor7	Factor8	Factor9	Factor10	Factor11	Factor12	Factor13
white sandstones	−0.5547	−0.1544	−0.1046	−0.4783	−0.0767	−0.1254	−0.1304	0.0415	−0.1122	0.0980	−0.0825	−0.1831	−0.1584
grey siltstones	−0.6026	−0.2647	−0.1722	−0.4867	−0.2243	−0.1103	−0.1885	0.0685	−0.0616	−0.0516	−0.0265	−0.1335	−0.0561
grey sandstones	0.0000	0.0000	0.0000	0.0000	0.0000	0.0000	0.0000	0.0000	0.0000	0.0000	0.0000	0.0000	0.0000
black shales	−0.3905	0.0931	−0.1227	−0.2419	−0.0966	−0.0582	−0.1216	−0.0892	−0.0213	0.2501	−0.0572	0.6381	−0.1146
green siltstones	−0.1747	−0.2121	−0.0850	−0.2062	−0.0595	−0.0437	−0.1804	−0.2174	−0.3218	0.1154	−0.0783	−0.3620	−0.0793
brown siltstones	−0.3867	−0.0953	−0.1109	−0.3672	−0.1125	−0.1022	−0.2346	0.1482	−0.0099	0.4772	0.1280	−0.0671	0.0845
red siltstones	0.0369	−0.0695	−0.0370	−0.0296	0.0174	−0.0327	0.0931	−0.0236	−0.1452	0.8315	−0.0839	0.0817	−0.0889
clay−ironstones	−0.3891	−0.1956	−0.2119	−0.2597	−0.1154	0.0341	0.0837	−0.0035	0.0669	0.0816	0.2429	−0.1489	−0.0814
other clastics	−0.2634	−0.0580	0.0429	−0.0835	−0.0029	−0.0967	−0.0448	0.6695	−0.0095	−0.1975	−0.1249	−0.1012	−0.1287
limestones and dolomites	−0.2648	0.1412	0.0214	0.2305	−0.0205	0.7086	−0.1096	−0.0701	−0.0033	0.1058	−0.1148	0.2019	−0.0179
chalk	0.3617	−0.1186	−0.0981	0.0080	−0.0532	0.7317	−0.0305	0.1643	−0.0788	−0.0671	0.0595	−0.1129	0.0008
chert	−0.1242	0.3245	−0.1332	−0.0367	0.2904	−0.0273	0.4715	0.0371	0.0408	0.3856	0.0973	−0.0698	0.1306
flint	0.0333	−0.0593	0.0778	−0.0774	0.0586	0.8664	0.0113	−0.0410	−0.0498	−0.0769	−0.0411	−0.0592	−0.0496
white granites	0.1063	−0.0427	0.8427	0.0547	−0.0234	−0.0546	−0.0628	0.0687	0.0747	−0.0280	0.0746	−0.0123	−0.0471
red granites	−0.0457	−0.1112	0.6765	0.2079	0.2136	0.3311	0.0648	−0.0043	0.1482	−0.2147	0.0763	0.0022	−0.0327
yellow granites	0.2732	−0.0453	0.7097	−0.0487	0.0712	−0.0103	0.1540	0.0164	0.0886	0.0269	0.2227	0.0929	0.0658
white/red granites	0.2083	−0.0494	0.5205	−0.0728	0.1587	−0.1629	−0.0233	0.2161	−0.3540	−0.0779	−0.3585	−0.1127	−0.0758
other granites	0.2342	−0.0518	0.4360	−0.0994	−0.0369	−0.0741	−0.0164	0.6748	−0.1853	0.0972	−0.1576	−0.0077	−0.1244
gabbro	−0.1785	0.0260	0.6197	0.5216	0.0028	0.0388	−0.0998	0.1139	0.0854	0.0444	−0.1085	−0.0912	0.0033
pyroxenites	−0.0328	0.0331	0.5943	0.3122	−0.0297	−0.0051	−0.1683	0.0677	0.0207	0.0017	0.2573	−0.0363	−0.0111
phyllites and mica schists	0.0833	0.0287	0.1998	0.2053	0.1579	−0.1303	−0.0425	0.0444	−0.1888	0.0242	0.5365	0.0507	−0.2221
serpentinites	0.1321	−0.0103	0.0087	−0.0133	−0.0753	0.1347	0.0165	0.7524	0.0265	0.1188	0.1066	0.0697	0.1287
white quartzites	−0.0051	−0.1070	0.1302	0.1767	0.7253	0.0086	0.1963	0.0220	−0.1244	−0.0249	0.0456	0.1960	−0.1714
grey quartzites	−0.1854	0.4880	−0.0524	0.0416	0.5970	−0.0077	0.1177	−0.0055	0.1140	0.0356	0.1422	0.1146	0.1750
green quartzites	−0.0974	0.5491	−0.1336	0.0330	0.5432	−0.0400	−0.0600	−0.0075	0.0028	0.1966	0.0671	−0.2315	0.1274
red quartzites	−0.0251	−0.1596	0.1092	0.2344	0.6498	−0.0371	0.0005	−0.0433	0.0070	−0.1245	−0.0696	0.0114	0.0864
other quartzites	0.0619	0.0608	0.0034	−0.0106	0.7597	0.0550	−0.0511	−0.0655	−0.0006	0.0717	−0.0578	−0.1630	0.0559
light gneiss	0.1526	−0.2586	0.1261	0.7315	0.2346	0.0784	−0.1099	−0.1475	−0.0088	−0.1904	0.1196	−0.0814	−0.0984
mafic gneiss	0.0268	−0.1776	0.1329	0.8588	0.1404	−0.0209	−0.0053	−0.0532	−0.0498	0.0213	0.1248	−0.0715	−0.0775
volcanic glass	−0.0253	0.5791	−0.0778	−0.0118	−0.0610	−0.0073	0.4300	−0.0169	−0.0363	−0.0414	−0.0315	−0.2438	−0.0898

Table 5.5. *cont.*

	Factor1	Factor2	Factor3	Factor4	Factor5	Factor6	Factor7	Factor8	Factor9	Factor10	Factor11	Factor12	Factor13
dark volcanics	−0.0350	0.9113	−0.0523	−0.0144	−0.0719	−0.0150	0.0630	−0.0330	−0.0030	−0.1069	−0.0444	0.0226	−0.0057
red volcanics	0.0670	0.8678	0.0968	−0.1589	0.0768	−0.0233	−0.0787	−0.0211	−0.0327	0.1516	0.0550	0.0359	0.0242
pumice	−0.0949	−0.0103	0.0089	−0.0438	0.0538	−0.0716	0.8923	−0.0142	0.0008	0.0358	−0.0073	0.0097	−0.0250
quartz	0.6780	−0.0060	−0.3079	0.1218	−0.0848	0.0185	0.4209	−0.0771	0.1402	−0.1240	−0.0961	0.1102	0.0856
white feldspar	0.7746	−0.0943	0.1466	0.0301	−0.2236	−0.0845	−0.2462	0.1200	0.0277	0.0657	−0.0181	−0.1389	−0.0148
red feldspar	0.8158	−0.1896	−0.0039	0.0034	0.0158	0.0959	−0.0876	−0.0456	0.1403	−0.0793	−0.0482	−0.1238	−0.0915
yellow feldspar	0.7760	−0.0407	0.1436	−0.1435	−0.0372	−0.0248	−0.1269	0.1133	0.0027	0.0960	0.1520	−0.1012	−0.1520
light green feldspar	−0.0094	0.0002	0.3743	0.0460	−0.1180	−0.0125	0.0333	−0.1150	−0.0144	−0.1027	0.6008	−0.0598	0.0883
garnet	0.0344	−0.0503	0.0028	−0.0295	0.0530	−0.1153	−0.1394	0.0255	0.6763	−0.1705	0.1450	0.0789	−0.1997
mafic minerals	0.0926	0.0693	0.0900	0.2502	−0.2494	−0.0672	0.1792	−0.0740	0.6045	0.0585	−0.1970	−0.2008	−0.0256
olivine	−0.0718	0.7464	−0.0930	−0.0793	−0.0933	−0.0114	−0.0407	−0.0441	−0.0045	−0.2076	−0.0388	0.3884	−0.0667
black opaques	0.0000	0.0000	0.0000	0.0000	0.0000	0.0000	0.0000	0.0000	0.0000	0.0000	0.0000	0.0000	0.0000
coal	−0.0659	−0.0117	−0.0200	−0.0783	0.1041	−0.0702	−0.0137	−0.0201	−0.0948	−0.0500	−0.0453	−0.0347	0.8754
biotite	0.2088	−0.0790	0.2119	−0.2015	0.0554	0.0207	0.0422	−0.0845	0.6438	−0.0412	−0.2609	0.0482	0.1052
muscovite	0.0000	0.0000	0.0000	0.0000	0.0000	0.0000	0.0000	0.0000	0.0000	0.0000	0.0000	0.0000	0.0000

Loadings of the variables on the 13 most impotant factors. Values for loadings vary from −1 to +1 and indicate how much of a certain lithology the factor contains. For example, a loading of 1 for gabbro would mean that the factor (source area) has the highest gabbro percentages found in the data set.

mainly of coarse crystalline granitic rocks. The highest percentages of feldspar were found in sample 23075, 20–39 cm core depth. This sample has a Late Weichselian age, is closest to the Norwegian continental margin of all analysed box core samples, and thus points directly to its source: the central part of western Norway. Therefore, strong positive scores of this factor indicate a sizeable contribution from western Norway.

Conversely, factor 1 has strong negative loadings for most clastic rocks, which means that the source contains little or none of those rocks. This general description fits well with the bedrock composition of western Norway. The highest factor scores (from 1.5 to 2.1) were obtained for Late Weichselian samples from the Norwegian Sea with very high proportions of feldspar and granitic rocks. A factor score is, mathematically, the projection of the samples' composition on the coordinate system formed by the factors. The greater the value, the more strongly the sample was influenced or supplied by the indicated factor, the source area.

An interesting aspect of factor 1 are the strongly negative loadings on clastic sedimentary rocks, in particular on grey siltstones (−0.6) and white sandstones (−0.56). The meaning becomes apparent when we look at samples with very negative scores (−1.7 to −2.7) on this factor: they include three samples from the Barents slope and several from the Norwegian Sea with very high concentrations of clastic sedimentary rocks, primarily grey silt- and white sandstones. Therefore, factor 1 has a double function: high positive scores indicate that the sample has received abundant IRD from western Norway, but when the scores are strongly negative, the debris came from the western Barents shelf.

Factor 2 has highest loadings on volcanic rocks, volcanic glass, olivine, and on grey and green quartzites. The 16 samples with highest scores of this factor (from 1.8 to 6.2) are located uniformly in the Greenland Sea, and invariably have a post-Pleistocene age. The percentages of volcanic rock IRD are highest in the vicinity of Jan Mayen, a small volcanic island in the Greenland Sea. Therefore, factor 2 represents IRD from Jan Mayen and IRD that is dropped in the course of the East Greenland Current, and which is partly of Arctic Ocean origin.

Factor 3 represents a plutonic source terrain with light and mafic intrusive rocks, including ultramafics. The top score of 5.2 is from sample 16052, 21–26 cm. This sample contains abundant clastic sedimentary rocks from the Barents Shelf, and it is thus possible that factor 3 represents plutonic rocks from Svalbard. But if that had been the case, the scores of this factor should be high for our samples from close to the Barents shelf, which they are not. Therefore, a Svalbardian origin is rejected and factor 3 probably represents rocks from northern Norway. But this is a somewhat tentative assignment.

The highest score of factor 6 was obtained for a Late Weichselian till sample from a coastal cliff facing the Baltic Sea in northern Germany (near Surendorf). The components are a thorough mixture of crystalline and sedimentary rocks derived from southern Sweden and the Baltic Sea (Boese, 1990; Kronborg et al., 1990). The highest loadings are on limestones and dolomites, chalk, and flint. Therefore, this factor represents IRD from southern Sweden and the Baltic Sea. The highest score is 3.47 for the Surendorf till with a combined total of 10.3% carbonates

Table 5.6. Age ranges during the Holocene during which coal is present

Core #	Age top	Age bottom	Coal
23269	0	1075	1.68
23071–2	0	1911	1.88
23060–2	0	6207	0.39
23053	0	9000	0.48
23055–2	0	10600	0.19
23070–2	191	1147	0.34
23073–2	195	2150	0.19
23069–2	205	2252	0.65
23072–2	215	2365	0.70
23068–2	246	2457	0.66
23057–1	900	9000	0.27
23070–2	1338	2867	1.18
23293	1505	2150	1.47
23071–2	2150	4778	45.83
23073–2	2345	3909	1.04
23069–2	2457	4914	1.01
23068–2	2703	6634	1.18
23070–2	3058	5733	6.60
23072–2	5160	6665	0.35
23067–2	5938	8190	6.90
23073–2	6255	8795	0.98
23068–2	6880	9829	0.42
23069–2	6962	8395	3.69
23072–2	8815	9890	0.11

Holocene samples with coal fragments and the respective age ranges. Left: Meteor samples, taken mostly during the summer of 1986. Right: Poseidon samples from 1987. Very long age ranges, for example 0–9000 or 900–9000 ybp, indicate samples in which the entire Holocene was taken as one single sample. This was generally the case when the Holocene thickness was 15 cm or less (box cores 23054–23057 and 23059–23066). Coal percentages are in the right column. Coal fragments were continuously transported to the Norwegian Sea. Samples younger than 2500 years bp contain less coal than samples prior to that date. Coal fragments are concentrated in the eastern Norwegian Sea in a corridor between the samples 23067 and 23073, between 0° and 3° East at 68.5 and 66° North. Interestingly, of all samples from the western Norwegian Sea (23059–23066), only one (23060–2) contained a fragment of coal.

Table 5.6. (cont).

Core #	Age top	Age bottom	Coal
16072	0	1200	3.33
16073	0	2250	1.62
16043	1176	2941	0.75
16072	1200	3600	2.93
16069	1462	3656	0.14
16073	2250	4500	1.47
16043	2941	4118	0.57
16072	3600	7500	2.46
16069	3656	5849	0.50
16052	4167	6667	0.57
16073	4500	6750	0.71
16043	5294	5882	1.78
16069	5849	7312	0.49
16043	5882	8235	0.37
16052	6667	10995	0.31
16073	6750	9000	1.08
16072	7500	9000	0.27
16043	8235	10000	1.03

and flint. The Norwegian Sea IRD sample with the next highest score of 2.13 is 23056, 17–24 cm, with an age close to the last glacial maximum. It has 2.9% carbonates and flint, which amounts to 28.2% of the original concentration of these lithologies in the till. Thus, the proportion of IRD from southern Scandinavia dropped to 30% and less because of advection of debris from other sources along the northward drift to the site of the Norwegian Sea core. These sources were the western and southern parts of Norway and perhaps also parts of the northern British Isles. The maximum proportion of IRD from southern Scandinavia during the last 450 ka was 30%.

5.4 THE HOLOCENE RECORD

The Holocene, the last approximate 10,000 years, is characterised by generally flat values for marine oxygen isotope ratios of foraminifera (Vogelsang, 1990; Voelker, 1999) and in the Greenland ice cores (Dansgaard et al., 1993; Mayewski et al., 1994), indicative of very little climate change or none at all during the ongoing interglacial. However, the IRD record of the Norwegian Sea is in stark contrast to the notion of climatic stability. As will be demonstrated throughout this chapter, the Norwegian Sea has not been entirely ice free. Instead, ice frequently drifted from places in the far north, between Svalbard and the East Siberian Sea, southward through the Norwegian Sea as it has done during the late stages of the last ice age (Bischof, 1994). These deep southward penetrations of Arctic ice (and surface waters) into the warmer, temperate regime of the Norwegian Sea are viewed as an occasional weakening of the North Atlantic Drift-Norwegian Current system. The Holocene ice drift suggests that the drift directions, and therefore the climatic setting of the Late Pleistocene, became reactivated occasionally during the Holocene. The Holocene climate oscillations occurred on two different time scales. The occasional southward penetrations of cold waters and ice from the north were brief and possibly occurred on decadal time scales. But on a somewhat longer time scale, the Holocene has undergone a change from initially vigorous ice rafting to very little in the middle, followed by a return to more ice rafting towards the present. This change also amounted to a shift of the southward currents that brought along the ice. The mid Holocene climate change occurred at 4200 years before the present and was probably global in nature. The evidence for these conclusions will be presented in this chapter.

The best examples of Holocene sedimentation come from the eastern Norwegian Sea (from 2° West to 5° East), where post-glacial sedimentation rates are higher than in the western parts. Some of the box cores from the Vøring-Plateau contain exclusively Holocene sediments and thus form the starting point of our treatment.

Before any interpretations can be made, however, the present physical setting of the Norwegian Sea must be understood. Despite its high northern location, the Norwegian Sea is ice-free throughout the year because it is kept relatively warm by the Norwegian Current, the northern extension of the North Atlantic Drift. Apart from a few rare occurrences, sea ice and icebergs are generally absent in the Norwegian Sea. In contrast, the Greenland Sea carries a seasonally variable sea ice cover with icebergs, and its waters are very cold, often close to the freezing point of seawater. The interesting question from a climatological point of view is whether this climate has been stable throughout the Holocene or whether it has changed.

The most important observation from this perspective is that the Holocene sediments contain coarse IRD, the maximum size of which is 16 mm. The IRD occurs loosely sprinkled throughout the sediment column in very low quantities. There is no discernible stratification. The Holocene grain concentrations of a few to a few tens of grains over 0.5 mm diameter per one litre sediment volume are several orders of magnitude lower than the very high Pleistocene densities of 5000 to over 30,000 grains per l sediment (Bischof, 1990).

Stratigraphically, the Holocene is clearly separated from the Pleistocene by its fine-grained character, abundance of marine microfossils, primarily foraminifera, and flat, unchanged oxygen isotope values (delta O-18,[15]) close to those of the present, approximately 2.3 per mil. The delta O-18 record does not reveal any climate changes during the last 9,000 years. But we will see that the IRD record underwent significant changes during the same time period, changes that will give us an insight into some of the most fundamental underlying principles of naturally occurring climate changes.

5.4.1 Distribution of coal fragments

The occurrence of coal in Holocene sediments in percentages as high as 45.8% is particularly intriguing, because it provides firm evidence that ice drifted from somewhere in northern Siberia, eastern Arctic Ocean, to the eastern Norwegian Sea, a region that is at the present ice-free all year. Identifying the source of coal in Holocene sediments is relatively easy, because coal is not exposed anywhere near the core sites in the Norwegian Greenland Sea at or above sea level. The closest source to the north is in Svalbard, but coal is also present further east in the sedimentary rocks of Franz Josef Land and in Siberia.

A counterargument to the proposition that the coal was indeed ice rafted and not the product of some other process was that coal also could have dropped from steamships cruising in the Norwegian Sea. A similar argument was used to suggest that some of the other dropstones could have fallen out from the stony ballast used by the Vikings to stabilise their boats. But both arguments are defused by the fact that, if true, they could only explain stones and coal fragments lying at the sediment surface. But dropstones and coal are found sprinkled throughout the entire Holocene sequence, and thus are in most cases clearly older than 1000 years. The sample with the maximum coal percentage of 45.8% is from 9–20 cm depth of core 23071, located at 67.5 North and 2 degrees East. The sample has an age of 4800 to 2100 years bp according to our age model.[14] Therefore, these coal fragments could not be remnants dropped by steamships or Vikings. Apart from samples of isotope stage 6 age, the Holocene samples with highest coal content (>3.7%) are from box cores 23067, 23069 and 23071, and from sub-surface levels between 9 and 41 cm depth. Their age ranges are 8200–5900 ybp, 6700–5100 ybp, 8400–6700 ybp, and 4800–2150 ybp. All other samples have less than 2% coal fragments. Of all 76 analysed Holocene samples (52 from Meteor and 24 from Poseidon cruises), 42 (55%) contained coal. The samples with coal are located in a wide area from 72.5° N and 10° West to 66° N and 3° East. Table 5.6 lists all Holocene samples that contain coal fragments, the ages of the samples, and the coal content in grain percent of the >0.5 mm IRD.

The possible sources of the coal can be Svalbard, where coal is currently mined, Franz Josef Land, and somewhere in Siberia. By virtue of our source area data set,[16] it is possible to compare the coal percentages of Norwegian Sea Holocene IRD to IRD from possible circum-Arctic sources, and to narrow the sources down to the most likely one. Three Holocene surface sediment samples (0–6 cm core depth) from northwest Spitsbergen represent IRD that is being dropped under the current flow of the West Spitsbergen Current. These samples are thus most likely to contain coal if the origin of the coal were Spitsbergen. But no coal fragments were present. This practically eliminates Spitsbergen as a source of coal. Of a total of 38 source samples from the Eurasian Arctic (between 30° West and 110° East north of 80°), only eight contained any coal at all, and of those samples only one exceeded 1%. This particular sample is located to the northwest of Franz Josef Land and thus could represent coal from there, but it is equally likely that the sample contains coal from further east that drifted to its place of deposition with the Transpolar Drift.[6] And indeed, by far the highest coal concentrations among ice rafted >250 μm grain assemblages were found in shallow marine sediment samples from the western part of the East Siberian Sea. Three samples off the mouth of the Indigirka River contained 9–30% coal fragments.[17] The Indigirka River drains a hinterland of Triassic, Permian and Carboniferous rocks which must be the primary host formations of the coal. Secondary sources of coal are the Yenisey River (one sample with 8.6% coal), the Ob River with 1.4%, and the Laptev Sea, where up to 1.6% coal are present. All things considered, the Indigirka River draining into the East Siberian Sea, and the Yenisey River are the primary sources of ice rafted coal for the Norwegian Sea during the Holocene. The coal was transported by drifting sea ice.

Ice drifted virtually continuously throughout the Holocene from the eastern Arctic Ocean into the Norwegian Sea, reaching the Norwegian continental margin. The exact drift path is difficult to reconstruct. Ice could have passed across the Barents Sea and then drifted southward to southwestward towards the target area between 0 and 3 degrees East in the Norwegian Sea. The ice could also have taken a different path, from the Barents Sea northwestward around Spitsbergen, then looping back southeastward with the East Greenland Current (this would also include ice coming directly from the Arctic Ocean through Fram Strait), and finally sweeping eastward somewhere between 65 and 70 degrees North towards our coal-bearing core sites. Of these two possibilities, the latter is less likely because it should have dropped more coal in the western part of the Norwegian Greenland Sea than in the east. Holocene sediments from the Greenland Sea (cores 16072, 16073 and 23269) contain coal, but in percentages lower than those occasionally achieved in the eastern Norwegian Sea. Furthermore, the relatively highest coal concentrations in the Greenland Sea occurred during the Late Holocene (2.9% from 3600–1200 years bp and 3.3% after 1200 years bp), whereas in the eastern Norwegian Sea, the highest values occurred earlier (three samples with up to 6.9% between 8400 and 3000 years bp). Therefore, a delivery of coal by drift ice first with the East Greenland Current and then swinging around eastward towards the eastern Norwegian Sea is possible for the youngest sediments which were deposited after 3000 ybp. But during the earlier part of the Holocene, the drift directions must have been different. Most likely, the ice drifted through the Barents Sea south- and southwestward towards the Norwegian Sea.

Figure 5.10 shows the possible ice drift directions during the Holocene for coal-bearing sea ice. Clearly, the reconstructed driftways deviate considerably from the present day oceanography and climatology of the region (Johannessen, 1986; Johannessen et al., 1994). At present, the eastern Norwegian Sea is under the influence of the northward moving Norwegian Current and its relatively temperate waters, and intrusions of ice from the north are impossible. Whether the appearance of sea ice from Svalbard and the eastern Arctic Ocean in that region means that the Norwegian Current weakened, thereby allowing cold, sea ice bearing currents from the Barents Sea (and ultimately the Arctic Ocean) to enter the area, or whether the ice was swept into the area by very strong, persistent winds against the reigning surface currents is not clear. What is clear, though, is that fundamental departures from the present-day climate situation occurred frequently during the Holocene, and that these departures were no isolated occurrences. Rather, these departures reveal some of the naturally occurring changes of the present-day climate in the Norwegian Greenland Sea and the Arctic Ocean, and point to a greater instability of the Holocene climate than previously thought.

In climate research, it is often difficult, if not impossible, to separate cause and effect. A major factor in the present oceanography of the study region is the North Atlantic Drift, which is mainly driven by westerly winds at low to mid latitudes. As

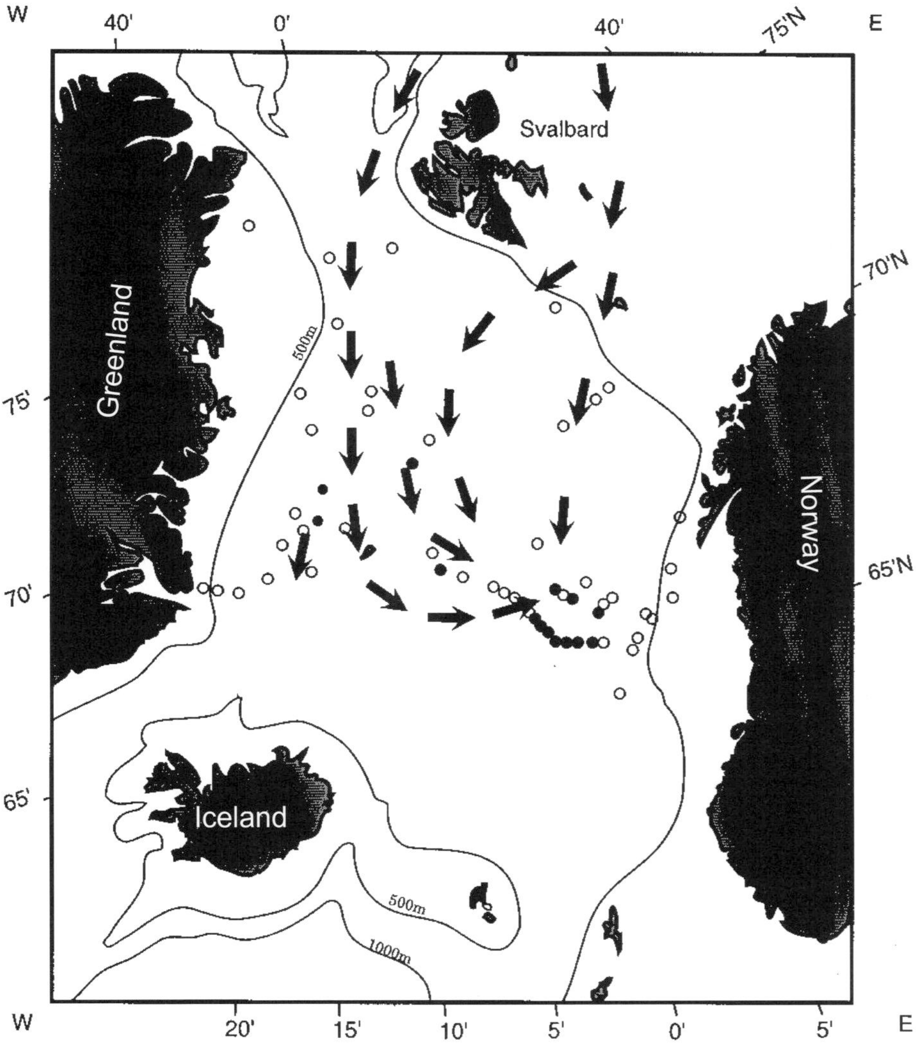

Fig. 5.10. Map of the Norwegian Sea with possible driftways of ice from the Arctic Ocean to the sample sites with coal fragments for the Holocene, <10 ka.

the waters flow to the north, they gradually lose heat to the atmosphere and cool. Since the density of seawater depends on its salt content and temperature, the cooling can make the water so cold that its density increases to the point where the surface waters become heavier than the waters they are surrounded by. As a consequence, they sink until they reach a level of equal density. The amount of sinking waters is compensated by a lateral inflow towards the vacated zone. This process is called deep water formation and occurs in different settings. Over the northern shelves, cold waters form in the winter and cascade down the continental slope. In the Greenland and Iceland seas, dense waters form when sea ice freezes from seawater. This process separates the surface water into two phases: sea ice with low salt content, and the remaining surface water with accordingly increased salt content. These waters are heavier not because their temperature changed, but because they received salt from brine rejection during the formation of sea ice.

If we apply this concept to our coal-laden sea ice drift, sinking waters over the Norwegian Shelf could conceivably pull surface waters and sea ice from the Greenland Sea eastwards into the Norwegian Sea. This could explain the sea ice with coal drift within the last 3–4000 years. But for the time before, the process must have been different. Cold waters perhaps form much more easily over the western Barents Shelf at a much more northern, and colder, location. If these waters sink, they can carry surface waters and ice from the Barents Sea southwestwards. This could explain the coal distribution during the earlier Holocene.

Another process is a changed atmospheric pressure distribution over the Arctic Ocean region. New modelling studies show that at present, the Arctic Ocean oscillates between two modes of circulation: one that keeps ice mostly within the Arctic Ocean in a gyral, circular circulation, and another one that allows massive amounts of Arctic pack ice to drift into the Greenland Sea (Johnson et al., 1999; Mysak and Venegas, 1998; Proshutinsky and Johnson, 1997). The latter style of the Arctic circulation resembles a strengthened and widened Transpolar Drift[6] and could have easily moved ice from northern Siberia (containing coal fragments) into the Norwegian Sea. Each of the two modes lasts for a few years before switching into the other mode.

But regardless of which process caused the sea ice-facilitated delivery of coal and, as we will see, other ice rafted lithologies, into the Norwegian Sea, it is important to consider the oceanographical and climatological consequences of such an expanded sea ice cover in the Norwegian Sea. The presence of sea ice introduces large amounts of freshwater equivalent into the otherwise normally saline (36.7 per mil) Norwegian Sea, which therefore becomes lower saline at the surface. The surface waters thus become specifically lighter and may no longer be heavy enough to sink when cooling occurs in the winter. Therefore, deep water formation may slow down or even come to a halt. And this, in turn, would slow the northward movement of temperate waters in the North Atlantic Drift[5] and initiate cold, sub-Arctic conditions in the northeastern Atlantic domain. Such events, followed by widespread cooling, are known as 'great salinity anomalies' (Häkkinen, 1993), and are attributed to above-normal quantities of ice leaving the Arctic Ocean.

Our coal findings in Holocene sediments of the Norwegian Sea provide firm evidence that ice drifted from a northern source (Siberia and Svalbard) to the southeastern Norwegian Sea. This is synonymous to a significant expansion of the sea ice margin in that region. Two main driftways are recognised: one through the Barents Sea to the south and southwest, and another from the Greenland Sea eastwards. Whether these intrusions were triggered by a changed atmospheric circulation over the Arctic Ocean or by intensified dense winter water formation over some of the shelves is not clear. But it seems that intensified bottom water formation over the outer Norwegian shelf could pull in surface waters and sea ice from the west (as observed), and that the same process over the western Barents shelf could trigger a surface flow from the northern Barents Sea to the south (also observed). Either way, a greatly expanded sea ice cover in the Norwegian Sea would result in widespread cooling over northwest Europe.

5.4.2 Holocene IRD grain densities

After deriving initial information on the directions of Holocene ice rafting from coal fragments, we now turn our attention to the IRD grain densities and the spatial distribution of IRD during the Holocene. Under present-day conditions, it is almost impossible for ice to reach this part of the Norwegian Sea for the following reasons:

1. At present, ice exists only a great distance from the core sites in the northern Barents Sea, around Spitsbergen, and in the Greenland Sea. Winter ice in the Baltic Sea melts in place and therefore does not reach the central Norwegian Sea. The fjords of western Norway are ice free all year round.
2. The surface current directions are strictly to the north, from warmer to colder areas. This motion and thermal regime would either melt incoming ice or deflect it into the Greenland Sea. Therefore, had these conditions existed uninterruptedly during the entire Holocene, not a single ice rafted grain should be present in the sediments. The fact that coarse IRD grains occur throughout the Holocene section means that ice frequently drifted into the Norwegian Sea, but where it came from and which drift direction the ice took must be determined from the composition of the IRD. The thickness of the Holocene section exceeds 45 cm in the area between cores 23066 and 23072, from 2° West to 5° East. The oxygen isotope curve in this section is flat at a value of 2.3 per mil (Vogelsang, 1990), indicating either minimal or no temperature and salinity changes in the Norwegian Current. In this area the upper 45 cm are younger than 8600 cal. years bp, and thus have an average sedimentation rate of at least 5.2 cm/ka. In all analysed cores the Holocene grain densities range from 0 and 80 grains per one litre sediment volume. These densities are 3 orders of magnitude lower than those of the Pleistocene which range from 5000 to more than 30,000 grains per one litre of sediment (Bischof, 1990). This 3 order of magnitude difference between Holocene and Pleistocene sediments is the same as in cores from the western Arctic Ocean (more about that in Chapter 6). The Holocene IRD grain densities uniformly show a temporal development from initially high to very low in the middle Holocene, before rising again to values comparable with the early Holocene.

The first example of grain densities is box core 23067–2, which contains a 40 cm thick Holocene (younger than 8600 years bp.; conventional beta-decay radiocarbon age transferred from 25 cm depth in core 23066–2) section. The core was split into five samples: 1–6 cm, 7–16 cm, 18–27 cm, 29–40 cm and 41–45 cm depth. The lowest grain counts were found in the middle Holocene, between 18 and 27 cm depth (Fig. 5.9). Lowest grain frequencies during the middle Holocene were also found in other cores from the vicinity of 23067–2 (Figs 5.11–13), and in X-ray photographs of all other cores from further west (core numbers 23059–23065; Bischof, 1990). The middle part of the Holocene was a time during which ice rarely entered this part of

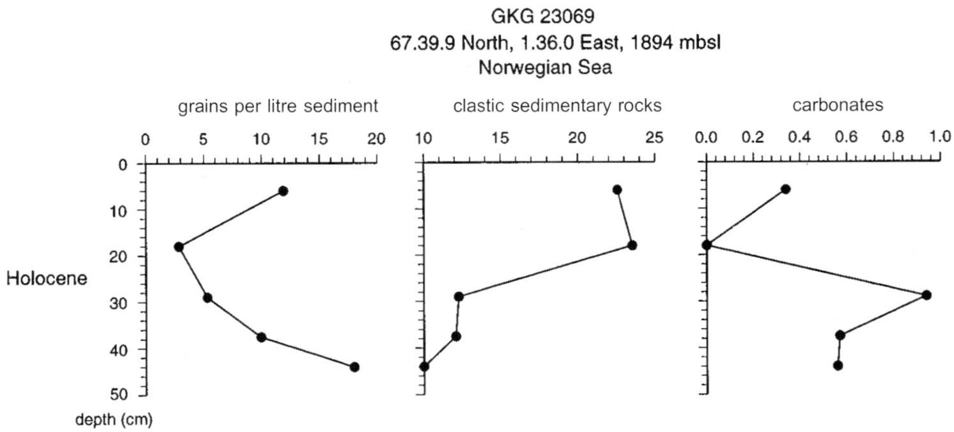

Fig. 5.11. IRD grain densities in core 23069 (left), along with the composition of the IRD (in grain percent), clastic sedimentary rocks and carbonates. Note the descent of grain densities during the early Holocene to a middle Holocene grain minimum at 18 cm, and subsequent rebound.

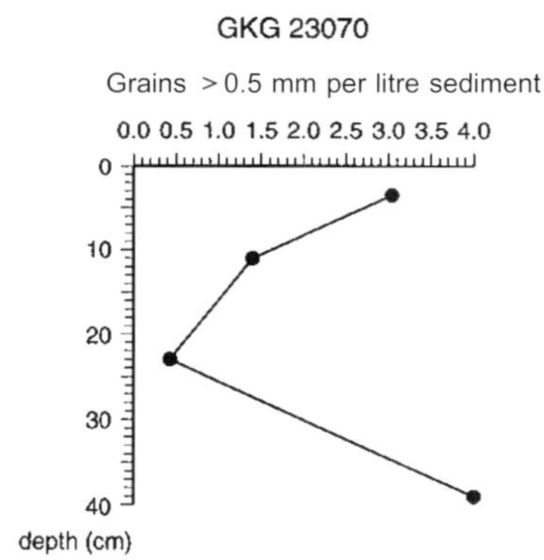

Fig. 5.12. IRD grain densities for core 23070. This core has the absolutely lowest grain concentrations of all Norwegian Sea stations. Compare the scale to Fig. 5.11: the maximum value of 23070 equals the minimum of 23069, and the highest values in 23069 are about five times higher than the maximum of 23070. The curve also descends to a middle Holocene minimum and rises again thereafter.

the Norwegian Sea. Earlier and later, ice rafting was more common. However, we must bear in mind that the Holocene IRD densities are only about one-thousandth of the Pleistocene values and, if this directly translates into the frequencies of icebergs and sea ice floes that drifted through this part of the Norwegian Sea in a given period of time, then the 'iceberg traffic' was only 0.001 as dense as during the Pleistocene.

Nevertheless, the fact that all four cores uniformly show the same initial descent, the middle Holocene low, and the Late Holocene rise, demonstrates that a profound,

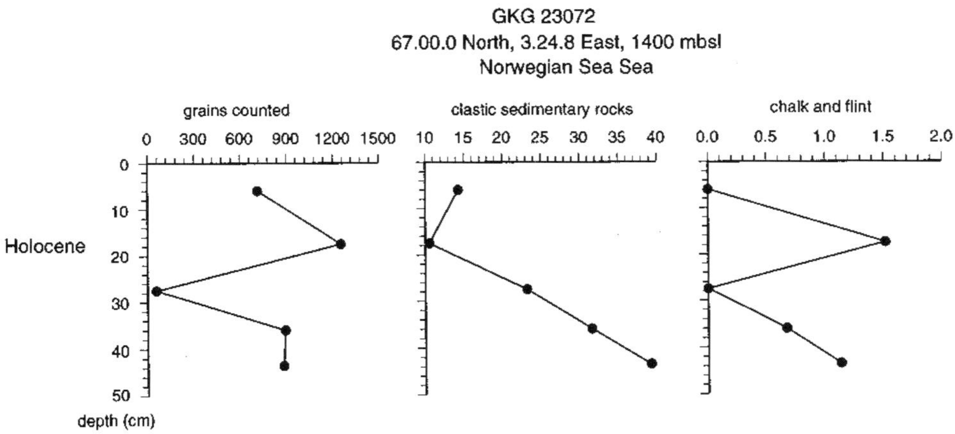

Fig. 5.13. IRD grain densities and composition in core 23072. Grain counts (left), and percentages of clastic rocks and chalk and flint. At this site, the percentages of clastic sedimentary rocks drop at the Middle Holocene grain minimum, as opposed to cores west of 23070 where this trend is reversed.

long-term change in the physical environment of the Norwegian Sea occurred during the Holocene. Two key points in time and space are the middle Holocene during which ice rafting was at its absolute minimum, and the location of core 23070 which has received lowest amounts of ice rafted debris during the Holocene. The Holocene climate, therefore, was not as steady and constant as the flat delta O-18 values suggest. But the IRD composition can provide the answer to the question about the nature of these climate swings.

5.4.3 Holocene IRD composition

During the early Holocene, IRD in core 23067 was dominated by crystalline lithologies and lesser amounts of clastic sedimentary rocks (Fig. 5.14a, b). The crystalline rocks were primarily light and dark coloured gneisses, quartz and feldspar grains, multicoloured quartzites, white and red granites, rare phyllites, mica schists, volcanic rocks, volcanic glass and pumice. Coal fragments are common, except during the deposition of fine grained middle Holocene, when there were none. Cherts and carbonates are very rare.

Therefore, the source areas of the Early Holocene IRD must contain widespread exposed granitic and medium to high grade metamorphic rocks in addition to lesser amounts of clastic sedimentary rocks. Occasionally the samples also contain chalk and flint fragments (Fig. 5.14). The occurrence of chalk and flint in this particular time period, long after the intense Pleistocene ice rafting was over, has important paleoclimatic implications. It demonstrates that ice from the receding southern margin of the Fennoscandian Ice Sheet drifted northward to a position of at least 67° N into the Norwegian Sea long after 8600 years bp. The ice could have been icebergs that broke off the ice sheet margin and then drifted through the marine channels

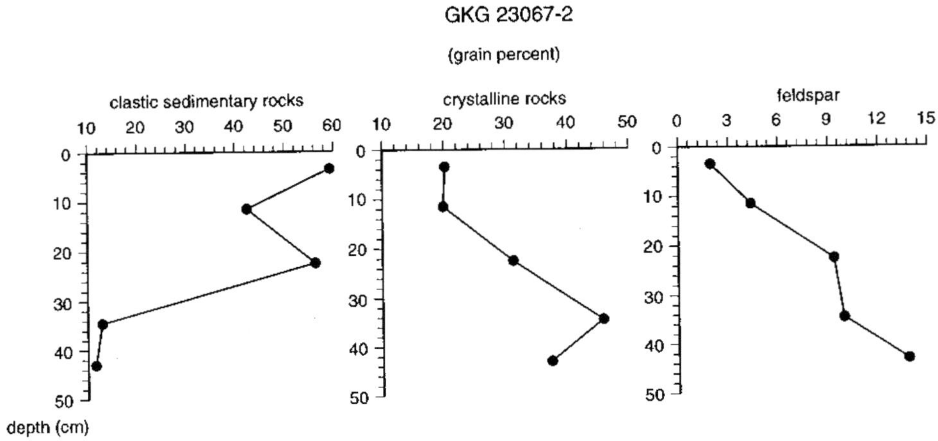

Fig. 5.14. Concentration of clastic sedimentary and crystalline rocks and feldspar grains in >0.5 mm IRD of core 23067. Note the early dominance of crystalline rocks and feldspar, which upsection gives way to abundant clastic rocks. Clearly, a major change in the predominant ice drift direction occurred during the middle Holocene, the period with lowest grain densities;

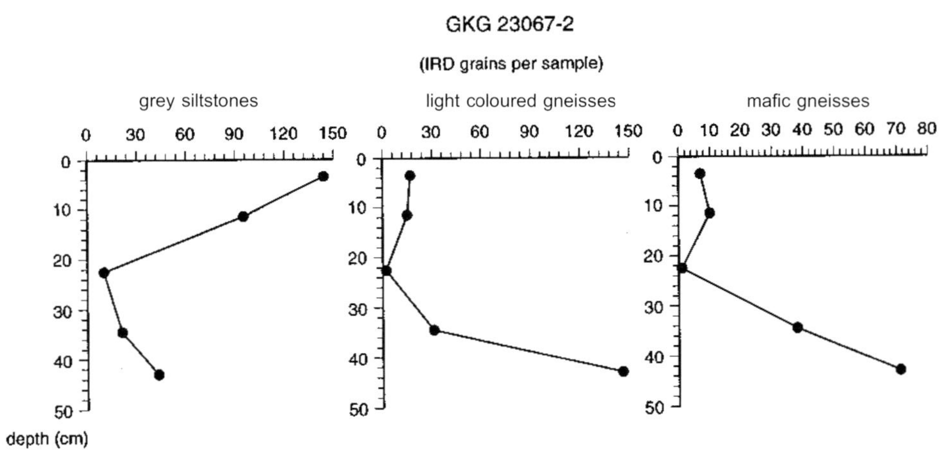

(b) In this diagram, the abundances of grey siltstones and light and dark coloured gneisses are shown as grains per sample. All grains in this core were counted, not just a split. The gneisses drop steeply during the early Holocene and do not rebound thereafter. By contrast, the siltstones also drop initially, but then rise to Holocene maximum abundances. Clearly, at this site, early Holocene ice rafting from a crystalline terrain has been replaced by ice rafting from a more clastic rock dominated terrain.

Fig. 5.14. (cont.)

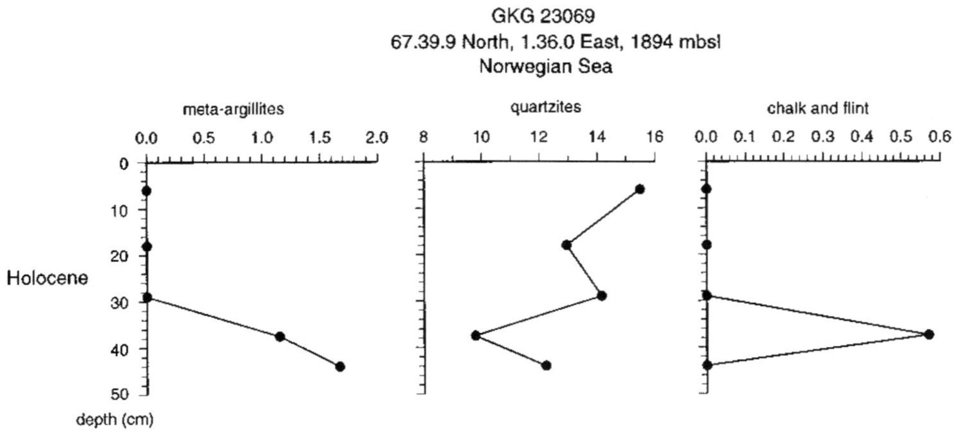

(c) IRD composition in grain percent for core 23069. Note the occurrence of meta-argillites and chalk and flint only during the Early Holocene, and the gradual rise of the quartzite proportion.

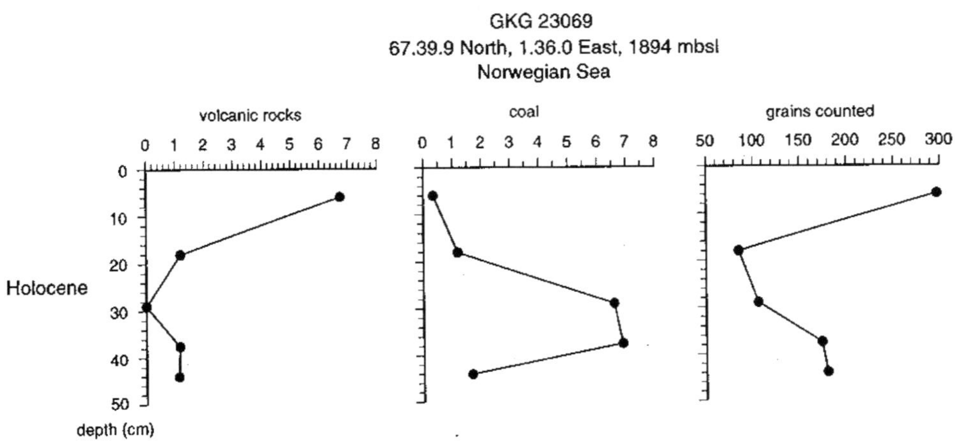

(d) Proportion of volcanic rocks and coal fragments in Holocene IRD of core 23069. The very high coal percentages during the Early to Middle Holocene denote a strong northerly influence from Svalbard or even the eastern Arctic Ocean, whereas the volcanic rocks indicate an increasing westerly influence from the vicinity of Jan Mayen during the more recent part of the Holocene. For control, the grain counts are also plotted on the right. None of the analysed samples contained more than 300 grains.

Fig. 5.14. (cont.)

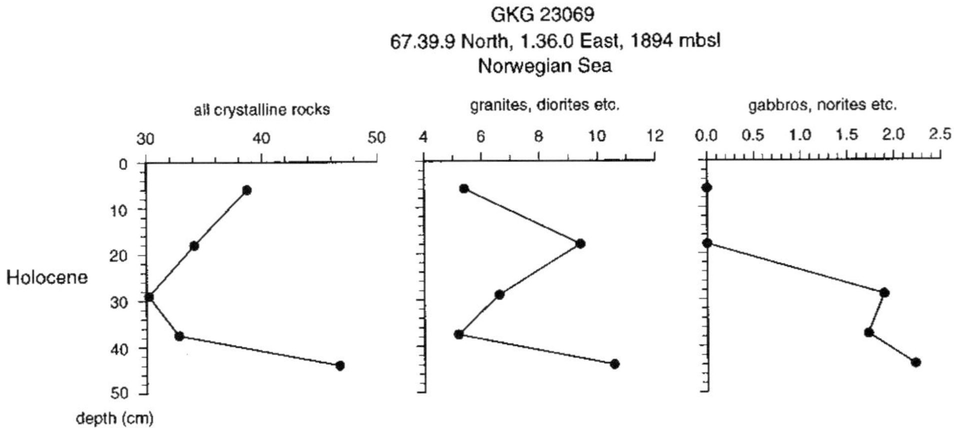

(e) This plot illustrates differences in the internal composition of the crystalline rocks. All crystalline rocks initially drop and then rise after a minimum of 30%. But the light and dark coloured intrusives differ from this pattern. Gabbros and norites are no longer part of the IRD inventory after the Middle Holocene. This proves that not just the overall abundance of crystalline rocks changed, but also that their origin was different.

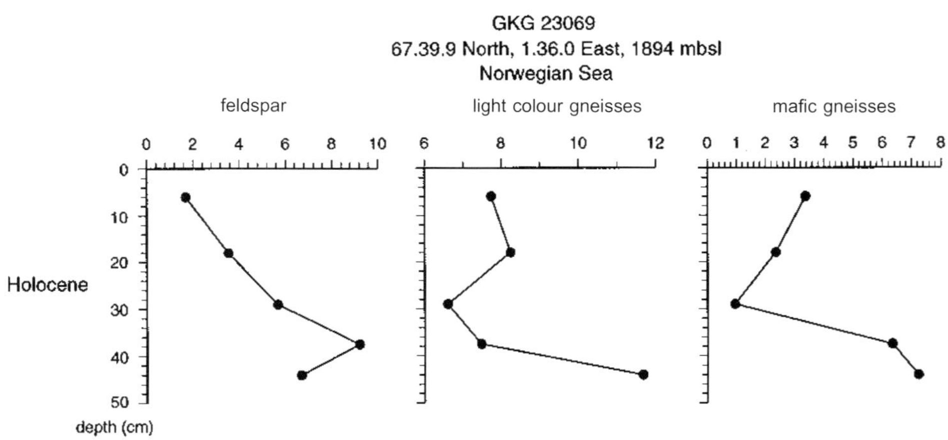

(f) Feldspar is in steady decline, very much like core 23067 (Figs 5.14a, b), and light and dark gneisses show an abundance similar to all crystallines with a minimum in the middle and rising values towards the bottom and top.

between Sweden, Denmark and southern Norway into the Norwegian Sea. Another possibility is rafting of winter and shorefast ice with inclusions of glacial till and/or pro-glacial sediments (Spjeldnaes, 1981). The southwestern Baltic region therefore was a source of ice to the Norwegian Sea during the early Holocene.

West of the location of core 23070, which has the lowest grain densities (= IRD content) of all Holocene samples in the Norwegian Sea, the IRD composition changes from crystalline rock-dominated during the Early Holocene to clastic rock-dominated during the Late Holocene (23067, Figs 5.14a, b). The shift is drastic, from 10% to values between 40 and 60%, an at least fourfold increase. And the shift occurs exactly during the fine-grained Middle Holocene interval. The same increase of clastic rocks is also seen in core 23066, where the clastic sedimentary rock percentages rise, from the bottom up, from 16.2 to 17.3 to 42.8%, before a slight drop to 32% at the top. Slightly further to the east, closer to the site of 23070, this tendency is still recognisable, but weaker (core 23069, Fig. 5.11). The clastic sedimentary rocks also rise at the middle Holocene IRD minimum, but not to 50–60% as in 23067. Here they only rise to 23%.

East of core 23070, this trend reverses. Here, in core 23072, clastic sedimentary rocks dominate first, during the Early Holocene, and then the composition shifts to multiple rock types (Fig. 5.13). Core 23072 contains significant amounts of IRD (see grain counts on the left of Fig. 5.13). A similar development from high amounts of clastic sedimentary rock IRD to a more mixed IRD assemblage thereafter is also seen in the Greenland Sea core 16073 (Fig. 5.15).

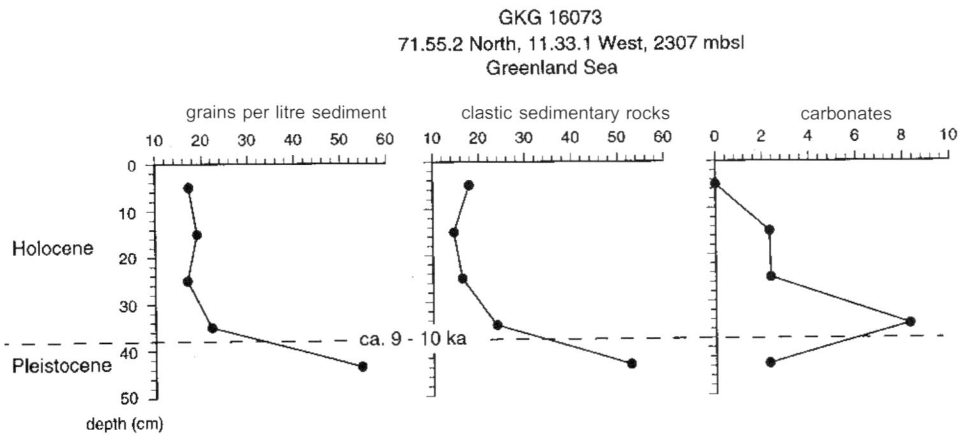

Fig. 5.15. (a) IRD concentration and composition (clastics and carbonates) of core 16073. The core is located on the flank of a submarine ridge (Vesteris Banken) in the Greenland Sea (Fig. 5.1). It contains mostly Holocene sediments, but the lowest sample is of Late Pleistocene age and has very high quantities of IRD. The Holocene–Pleistocene boundary is at 38 cm depth. The plotted value of 55 is based on the number of grains in the counted split, which was only a very small fraction of the whole IRD in this sample. But the split was not determined, and therefore this data point is artificially low. In contrast to the Norwegian Sea, where the middle Holocene has lowest IRD densities and the early and late stages have more, there is no discernible difference here. Rather, the IRD grain densities stay flat at a level of 20 grains per litre sediment. The IRD is initially high in clastic rocks, similar to the eastern Norwegian Sea close to Norway, but during most of the Holocene, clastics are at the 20% level. Carbonates are high during the early Holocene and drop thereafter. For carbonate percentages of 8%, a contribution from the Arctic Ocean (North Greenland, Svalbard, the Canadian Arctic Archipelago) is necessary.

Fig. 5.15. (cont.)

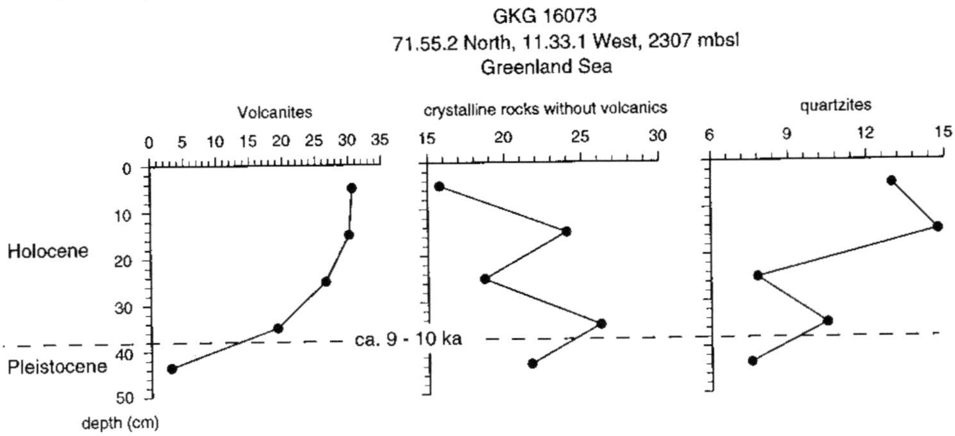

(b) Volcanic rocks are increasing after the Pleistocene to values of 30%. Values this high are only found in this part of the Greenland Sea in the vicinity of Jan Mayen, a small volcanic island. The volcanic rocks form a characteristic assemblage of grey and black, fresh looking basalt, brown and black volcanic glass, pumice, and bottle-green olivine grains. This type of volcanic assemblage is diagnostic for Jan Mayen and represents sea ice rafted debris. It is only found within a 200 km radius around Jan Mayen.

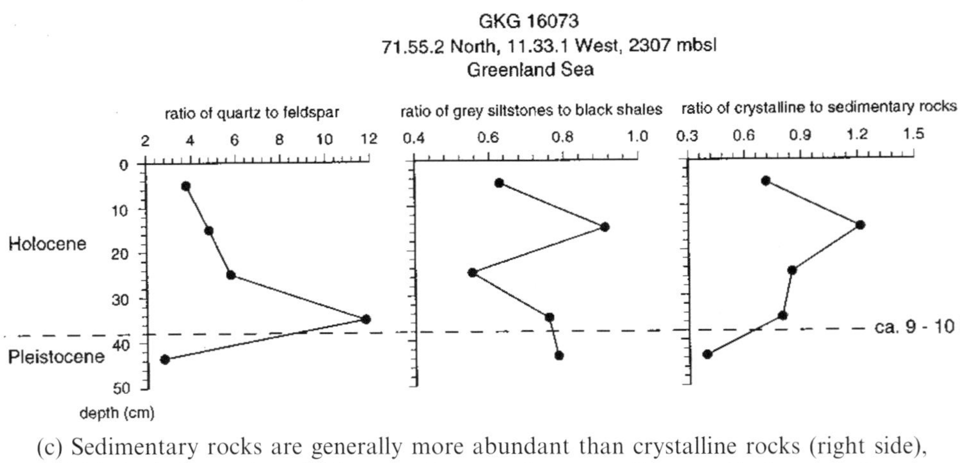

(c) Sedimentary rocks are generally more abundant than crystalline rocks (right side), and black shales are invariably more frequent than grey siltstones. Both properties are characteristic for IRD that is transported by ice in the East Greenland Current. The origin of this IRD is mostly in North Greenland and Svalbard.

The most pronounced change of the IRD composition during the Holocene occurred during its middle part: a shift between clastic sedimentary and crystalline rocks. This change was related to changing sources of ice and icebergs, and to changes of the surface circulation. In the following section, both aspects of the IRD, its relative abundance in the sediment and its composition, will be integrated into a model of the Holocene surface circulation.

5.4.4 Holocene ice drift and surface circulation

The most important observations with respect to the Holocene are that

(a) ice episodically drifted into the Norwegian Sea,
(b) there was an early and late Holocene ice rafting maximum,
(c) the IRD composition changed during the middle Holocene minimum,
(d) the IRD sources were mostly in the north, but also occasionally in the south (chalk and flint in 23072!),
(e) core 23070 has the overall lowest amounts of IRD, and
(f) that the IRD composition was different east and west of the site of core 23070.

All these observations are best explained by the competing effects of two major forces: the warming force of a warm to temperate, northward flowing Norwegian Current (NC) without any ice, and the cooling force of cold, southward flowing, sea ice and iceberg-laden currents. Obviously the area of maximum strenght of the NC is the site of core 23070, which has the lowest IRD densities from anywhere in the Norwegian Sea. The centre of the warm NC thus is at the zero meridian, more than 500 km off the coast of Norway.

To both sides of the NC, colder and occasionally ice-bearing currents could establish themselves. These currents were most probably extensions of cold currents that still exist today, but in different parts of the Norwegian Greenland Sea. One cold current (the East Spitsbergen Current) at the moment flows southwestwards around the east side of the Svalbard archipelago in the Barents Sea, northward up along the west of Spitsbergen, and then loops back to the south, joining the cold East Greenland Current. At the present time, the East Spitsbergen Current carries ice and icebergs from around Svalbard and the eastern Arctic Ocean, and the East Greenland Current carries ice from the Arctic Ocean. In the past these two cold currents – which presently flow together in the Greenland Sea – were frequently split by the northward flowing NC. During those times, the East Spitsbergen Current was significantly extended and flowed southwards into the Norwegian Sea, bringing along icebergs and sea ice with coal fragments as seen in core 23069. But there were also times when northward moving currents on the right hand side of the NC brought ice (winter ice, shorefast ice, anchor ice?) from as far south as the western Baltic Sea into the Norwegian Sea (as seen in the chalk and flint content of Early Holocene IRD). The waters of this current, a version of the Norwegian Coastal Current (NCC), were colder and less saline than the NC because of the less saline waters of the Baltic Sea and river runoff and meltwaters from southern Sweden and Norway, and thus facilitated the long ice drift path. The ice rafted particles dropped to the sea floor when the ice and icebergs that carried them melted upon contact with the warmer waters of the NC.

Because the shells of foraminifera that lived in the NC show virtually no change of their water temperature or salinity, the observed changes of the IRD density and composition are explained by the different strength and geographic extent of the NC rather than by a change of its temperature or salt content. During the Early Holocene, ice rafting was fairly intensive, and the ice came from both southern and northern sources. During this period, the NC was relatively weak and allowed southward drift of ice and icebergs from northern Spitsbergen, where mostly crystalline rocks occur on its left side (west), and southward currents from southern Spitsbergen, where clastic sedimentary rocks are more frequent, on its right (east). On this same right side, northward moving currents with ice from the Baltic region also occasionally established themselves. This situation is depicted in Fig. 5.16a.

During the Middle Holocene, ice rafting was minimal and a broad, strong NC established itself (Fig. 5.16). It flowed northward on a wide track across the central Norwegian Greenland Sea parallel to the zero meridian. Its strength pushed the cold, ice-laden currents back to far northern locations. During this period, the Norwegian Sea was truly ice-free.

Later, a return to conditions similar to those during the Early Holocene occurred (Fig. 5.16). The NC is pushed southwards by two cold currents with ice and icebergs. To the left of the NC was a current from southern Spitsbergen, and to the right side a current from northern Spitsbergen. The southern influences were weaker than during the early Holocene.

The East Greenland Current existed more or less unchanged during the Holocene. Changes in the IRD composition were largely due to different areas of sea ice and iceberg formation rather than changes in the current direction.

We have seen that two types of climate change have occurred during the Holocene. One is an expansion of cold currents with ice and icebergs into the Norwegian Sea. These events were probably of short duration and symbolise the short-term, naturally-occurring variability of the atmospheric and surface ocean circulation of the Nordic Seas. But there is also a long-term trend from the Early to the Middle Holocene to less ice rafting, with an absolute minimum during the Middle Holocene, and then a return to increasingly intensive ice rafting towards the present day. The latter observation is particularly intriguing, because the IRD record (grain densities) looks very much like a return to colder, possibly ice age, conditions. Gerard Bond, a paleoclimatologist at the Lamont Doherty Earth Observatory, gave a talk at the annual fall meeting of the American Geophysical Union in San Francisco in 1998 in which he described an IRD composition change towards the present that looked remarkably similar to an older change that was followed immediately by the Wisconsinan (in Europe: Weichselian) ice age. The important implication from this is that, if this change in the IRD composition, and thus a change in the direction of ice rafting, initiated the last ice age, the most recent change may be the harbinger of the next ice age. Therefore, we may have to entertain the possibility that Earth's natural climate development may be on a return to another ice age, or at least to colder conditions than at the present. Ironically, the expected global warming because of anthropogenic release of greenhouse gases may cancel out this effect.

Without a doubt, the long-term changes in the Norwegian Sea IRD record make the Holocene look more like an intermission between two consecutive ice ages, with the next one to begin at any time. The warmest period, and the one with the most pronounced IRD composition change, was the Middle Holocene. In our cores, this interval occurs between 18 and 27 cm core depth. The average of five measurements

Fig. 5.16. Ice drift during the Holocene.

(a)

(b)

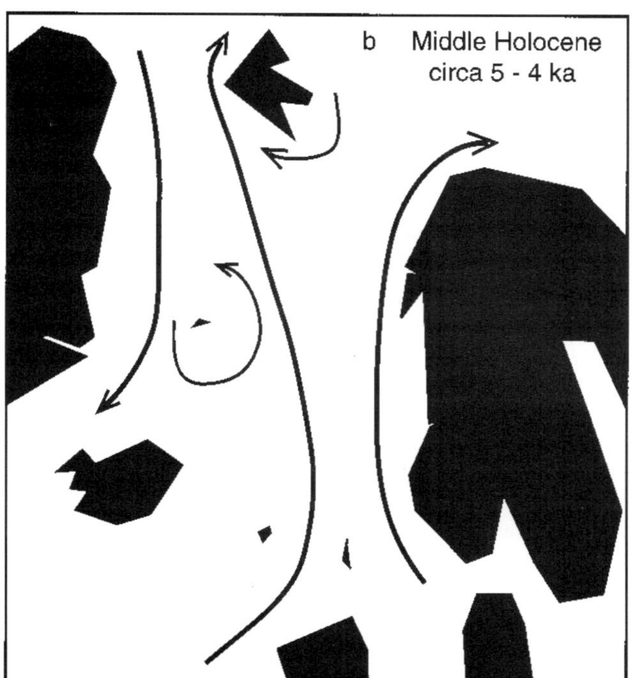

Fig 5.16. (cont.)

(c)

(a) The Early Holocene situation had a Norwegian Current in the centre that split two cold currents in half. On the right side was an extended version of the East Spitsbergen Current with ice and clastic sedimentary rock IRD from southern Svalbard. On the left was a current from western and northern Spitsbergen with mostly crystalline rock fragments. Between these opposing currents, gyres (shown by circular arrows) developed in which warmer and colder waters mixed. These mixing zones received most IRD because of melting of sea ice and icebergs upon encountering warmer waters. A branch of the Norwegian Coastal Current carried ice from the Baltic Sea into the Norwegian Sea (arrow at the lower right).
(b) During the Middle Holocene, ice rafting was least intense. The NC was at its strongest and broadest, moving northward through the centre of the Norwegian Sea, accompanied by the Norwegian Coastal Current to the east. This scenario kept the Norwegian Sea mostly ice free, and cold, ice-carrying currents were restricted to the north.
(c) The Late Holocene saw an intensification of ice rafting. Again, as with the Early Holocene, cold currents with ice and icebergs drifted southward on both sides of an NC which was pushed back and of less geographic extent than before. The main differences to the early Holocene are (i) that the sources of ice for the eastern Norwegian Sea shifted to North Spitsbergen (crystalline rocks), (ii) the sources for the central Norwegian Greenland Sea shifted to southern Spitsbergen (clastic sedimentary rocks), and (iii) the influences from the south became weaker (broken line in the lower right).

is 22.5 cm. At an average sedimentation rate of 5.2 cm/kyr (calculated with 45 cm and 8600 years), the age of the period of maximum northward extension of the NC is 4330 years before the present. At almost the same time, 4200 cal. years bp, the Akkadian empire, the first highly developed civilisation in northeastern Syria, vanished without any indications of violent interference for a period of 500 years (Cullen et al., 1997). During the same time period, the sediment composition at the archeological excavation site and in cores from the Arabian Sea changed from a fertile soil to a dry, salty, gypsum-containing desert dust. Therefore, the demise of the Akkadians is most probably the consequence of a change of the prevailing wind directions in the Middle East. But these changes in the atmospheric and oceanic circulation may not have been restricted to the northern hemisphere: Lamy et al. (1999) describe a change of the prevailing wind directions and fluvial discharge in the southern Pacific Ocean off the coast of Chile at about the same time, 4 ka.

I herewith suggest that these near-synchronous changes of the surface ocean circulation in the Norwegian Sea and the atmospheric circulation over the Middle East and the south Pacific were no coincidence. They represent profound climate changes within the Holocene that affected at least the northern hemisphere, if not the entire globe. Obviously, this change was a mixed blessing: what may have created relatively pleasant living conditions in northern Europe was a death sentence for the continental Middle East.

5.5 THE WEICHSELIAN GLACIATION IN THE NORWEGIAN SEA

In the following chapters on the Weichselian glaciation in the Norwegian Greenland Sea (NGS) it will be demonstrated that this ocean was influenced by two different surface current regimes: one with a dominantly southward flow in the Norwegian Sea, and another one with northward flow. The Greenland Sea was under the influence of southward currents throughout the entire Pleistocene. Between the northward currents in the east and the southward currents in the west, a counterclockwise rotating circulation developed in the central NGS where IRD lithologies from many different places were mixed together. From the composition and lithologic diversity of the IRD it will become clear that icebergs were fed from different parts of the Late Weichselian ice sheets into the changing surface currents. Distinct episodes of massive iceberg calving alternated with quieter periods during which ice rafting was minimal. During every iceberg rafting episode, the icebergs came from different sources.

The eastern Norwegian Sea was supplied by icebergs from the southern and western parts of the Fennoscandian Ice Sheet (FIS). Towards the west, the IRD came increasingly from northern Norway and Svalbard. Further west, in the central NGS, an increasing portion of IRD from North Greenland and circum-Arctic sources is noticeable.

The driving forces of the surface ocean circulation were (a) the atmospheric pressure distribution over the Arctic domain, (b) the strength of the inflow of North Atlantic waters through the southeastern Norwegian Sea, and (c) the amount of freshwater absorbed into the northwestward flowing North Atlantic waters. Intricate feedbacks exist between the atmospheric circulation and ice sheet growth: the atmosphere controls ice sheet growth through the direction of moisture supply, and in turn the growing ice sheets changed the atmospheric circulation owing to the formation of strong and stable high pressure cells over them.

Iceberg calving from the Pleistocene northern hemisphere ice sheets was sequential, not synchronous. Episodes of massive iceberg calving in distinct sections of the ice sheets were not only triggered by climate changes as seen in the Greenland ice, but also by internal ice sheet oscillations.

The evidence for all these conclusions will be laid out in the following sections.

Glacier growth in northern Europe began at about 120,000 years bp, becoming increasingly stronger and finally leading to widespread continental glaciation, the last ice age. The gradual, long-term buildup was punctuated by brief periods during which large quantities of glacial icebergs were released into the Norwegian Sea. Baumann et al. (1995) calculated the accumulation rates of coarse-grained ice rafted debris and identified two main phases of the Weichselian Glaciation: a first, minor stage from 120–70 ka with limited ice extension to the Norwegian coastline and relatively little ice rafted debris, and a second phase from 70–10 ka during which ice frequently advanced onto the shelf. Within this latter phase there was continuous delivery of ice rafted debris. Baumann et al. (1995) recognised ten peaks of maximum accumulation rates of coarse IRD (in grams per $cm^2/10^3$ years). The stop-and-go character of iceberg discharges during the last glaciation is further confirmed by an ultra-high resolution textural analysis of a sediment core from close to the Norwegian continental margin with very high sedimentation rates. The study (Voelker, 1999; with data obtained by Vogelsang, 1999) clearly reveals sharp, outstanding IRD peaks which are separated from each other by fine-grained sediments. These IRD peaks could be analogue events to the so-called Heinrich Layers in the North Atlantic (Bond and Lotti, 1995), the product of episodically occurring massive discharges of large amounts of icebergs from the Pleistocene ice sheets.

5.5.1 The record of KAL 23074

The premier example of Late Weichselian ice rafting is sediment core KAL 23074. This kastenlot core has the highest sedimentation rates of all cores recovered during the Meteor II cruise in 1986 (Gerlach et al., 1986), and offers the unique opportunity to study the Weichselian Glaciation in a vastly expanded sedimentary record. Of all cores from the Meteor II cruise, KAL 23074 is located closest to the Norwegian continental margin (66°40.4N, 4°54.4E, 1157 mbsl; Fig. 5.1). Because of its proximity to Norway, it serves as a premier example of IRD derived mainly from there, possibly with some debris from other sources. Changes in the composition of the IRD reveal changes in the icebergs' origin and thus changes in the prevailing ice drift directions. The closest distance from the core site of 23074 to the Norwegian coast is 370 km, and its distance to the shelf-break, the site of the maximum Late Weichselian ice advance, is approximately 100 km. This location is important, because it means that even during times of maximum ice advance onto the Norwegian shelf, sometime between 26 and 20 ka, icebergs still had to drift at least

100 km offshore to reach this site. Before 26 ka and after 20 ka, when the ice margin was located near the Norwegian coastline, this distance was even greater, approximately 400 km. And this is the drift distance for icebergs from the nearest source in coastal western Norway. As we have already seen in the former, more general treatment of the Norwegian Sea IRD composition, the IRD is mixed and consists of components from many different locations, not just the ones closest to the core sites. It will be demonstrated that the length of the iceberg driftpaths during the Pleistocene were much longer than the direct distance to the closest IRD source, in many cases exceeding 1000 km. But these are average vector distances that assume a straight drift. In reality, however, the drift paths of sea ice and icebergs can be crooked or resemble a 'drunkard's walk', as Roger Colony once called it (Colony and Thorndike, 1985). Also, ice and icebergs can be trapped in a circular loop. This happened to ice island T-3, which was used as a drifting research station in the Arctic Ocean during the 1970s. This ice island broke off a glacier in northern Ellesmere Island and completed three clockwise turns with the Beaufort Gyre in the western Arctic Ocean before drifting out through Fram Strait. During this drift, nearly 600 sediment cores were taken, which formed our sole source of information about the sediments in the Arctic Ocean up to the mid-1980s. These cores are described in the chapters on ice drift in the Arctic Ocean. The first comprehensive presentation of results from the T-3 station can be found in Clark et al. (1980). Therefore, even if a net drift path of 1000 km or more is identified by the IRD record, the actual drift length of the icebergs may have been much longer.

Core 23074 contains IRD that was deposited between 25 and 10 ka.[14b] The oxygen isotope values (Fig. 5.17a) are generally heavy between 700 and 130 cm core depth, equivalent to an age range of 30,590 to 18,200 ka (Voelker, 1999; Vogelsang, 1990), indicating cold and saline surface waters, characteristics of full glacial conditions. Only the lower part of the core is punctuated by short excursions to the right (lighter values), which indicate meltwater from the ice sheets or melting icebergs, or climatic warming (possibly similar in style to the Dansgaard/Oeschger events recorded in Greenland ice), or a combination of both. Interestingly, the IRD grain densities (Fig. 5.17b) clearly contradict a commonly held belief that periods of massive iceberg discharges which produce the very coarse layers only occur during deglaciations. They are therefore believed to be synchronous with light oxygen isotope values. But the 23074 curve (Fig. 5.17b) shows multiple, very massive, IRD peaks during a long period of consistantly heavy O-18 values. Undeniably, these iceberg discharges occurred during very cold intervals. They were neither caused by, nor have they triggered, a climate change. Instead, they were most probably the product of internal ice sheet oscillations such as those envisioned by MacAyeal (1993).[18]

5.5.1.1 Stratigraphic position of the samples

The eleven samples analysed for IRD composition from core 23074 represent very different climate scenarios: the top three samples are from the final rise of the oxygen isotope curve to values of the present. Sample 3 (the third from the top and the lowest sample from the GKG, giant box core) is from just before the last rise of oxygen isotope values when these values were still relatively heavy and IRD grain

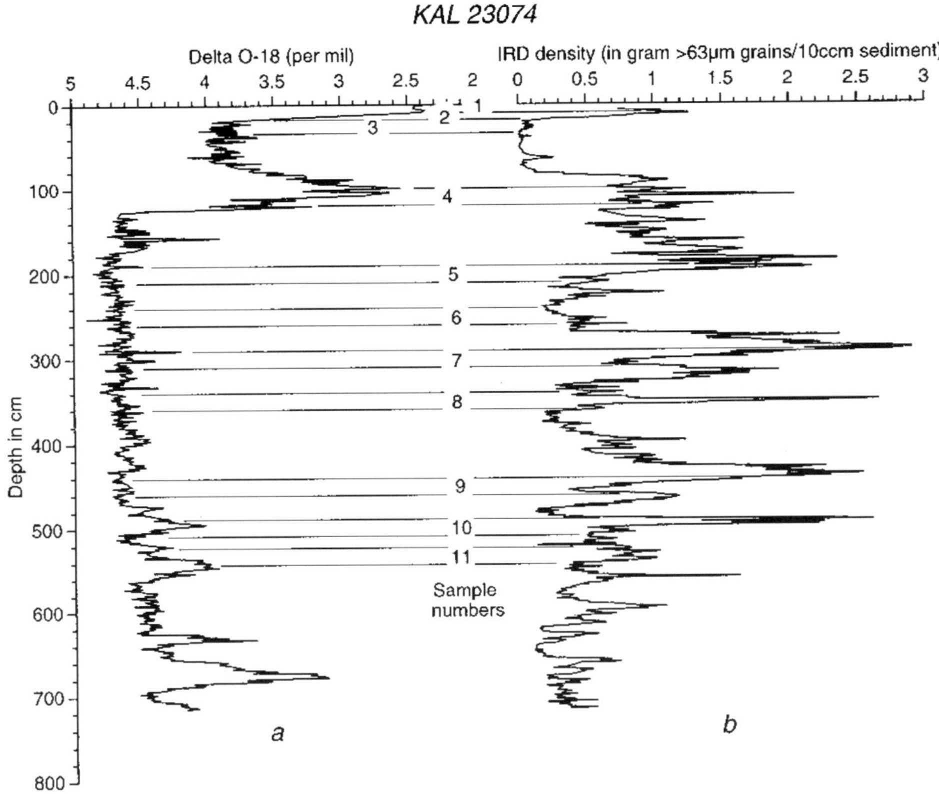

Fig. 5.17. Delta O-18 values (a) and IRD grain densities (b) versus depth for kastenlot core 23074 from the Mid-Norwegian continental slope. IRD densities are given as weight of grains over 63 µm (in grams) per sediment volume unit (10 cm^3). Note the drastic oscillations of the IRD densities between low background values of 0.1–0.2 and maxima near 3. The stratigraphic positions of samples analysed for IRD composition are indicated by fine horizontal lines, with the sample numbers (from 1 to 11) in the centre. The IRD samples represent an age range from 25 to 10 ka. The samples represent very different stages of the last glaciation: sample 11 represents the composition of IRD in a minor peak, sample 10 is the composition in a massive IRD peak at 500 cm core depth, and sample 9 represents the lower flank of the next massive peak upsection. Sample 8 is another very coarse unit, and sample 7, similar to sample 9, is from the lower flank (resp. the beginning) of another very coarse peak at 290 cm depth. Sample 6 represents a zone with relatively low IRD grain concentrations. Samples 5 and 4 are from the beginning and the near end of the thickest coarse layer in this core. One of the most drastic changes of the IRD composition, paleo ice drift and surface current directions, took place during the interval between these two samples, some time between 19 and 18 ka.

densities were low. Sample 2 is from the final rise in O-18 values, which in this case is accompanied by a rise in the IRD grain densities as well. The top sample (1) represents the end of Pleistocene ice rafting at about 10 ka. Samples 4 and 5 represent the beginning and the near end of the thickest (130 cm) coarse section in this core. Sample 5 represents the lower flank of this coarse interval and is from one of the coldest stages of the last glaciation with very heavy delta O-18 values of almost 5. Coarse-grained ice rafted deposition continued at variably high levels for the next 60 cm until sample 4 which is from the first rise of oxygen isotope values after the last glacial maximum. This period is also sometimes referred to as the first step of the last deglaciation, or Termination Ia. One of the most drastic changes in IRD composition occurred in the time between the deposition of samples 4 and 5, that is, between 19 and 18 ka. This change is a dramatic replacement of a large portion of the crystalline rocks, quartz and feldspar fragments which dominated the IRD before 19 ka by a lithologically uniform, low diverse assemblage of clastic sedimentary rocks from the Barents shelf. This event will be discussed in more detail in Section 5.8.

Sample 6 is from a similarly heavy and cold, but not quite as coarse interval as sample 5. Sample 6 may therefore be regarded as an example of glacial background sedimentation during times when low to moderate amounts of icebergs drifted in very cold and ice-covered waters, and only small quantities of coarse IRD dropped to the sea floor. Sample 7 is from the lower flank of a coarse IRD peak at 280–320 cm depth. Sample 8 represents a thin but very coarse IRD peak at 350 cm depth in its entirety, while sample 9 is an analogue to sample 7, from the lower flank of a coarse peak centred at 440 cm. Sample 10 represents a small meltwater and large IRD peak at 500 cm, and sample 11 was taken from the upper slope of a meltwater peak at 540 cm. Here, the meltwater peak is followed by an IRD peak with moderate grain concentrations.

Since sampling for analyses of the IRD composition focussed on the coarsest intervals, and because 20 cm-thick slices of sediment were taken in samples 4 through 11, the IRD compositions obtained for these samples are the result of integration over 20-cm thick core sections. But because the densities of coarse-grained ice rafted debris (reported as gram >63 μm per 10 cm^3 sediment) changed during some of the sampled intervals, the composition is biased towards the coarser sub-sections within the sampled intervals. For example, the IRD composition of sample 4 will be more influenced by the sharp, very coarse peak in the centre of the sample interval at circa 110 cm rather than by the debris above and below. Similarly, the composition of sample 5 is biased towards the top of the sampled interval from 210–190 cm. This is because the grain concentrations are 4–5 times higher at the top (at 190 cm) than at the bottom (0.3–0.7 g >63 μm per 10 cm^3 sediment at the bottom versus 2.2 at the top). Similar considerations apply to samples 7 and 9 which, similarly to sample 5, represent the lower flank of major IRD peaks and thus are biased towards the upper, coarser, portion.

Here is a short list of what the samples represent:

1. the end of Late Weichselian ice rafting;
2. termination Ib;

3. last segment of the Younger Dryas;
4. the coarsest peak in Termination Ia;
5. massive IRD peak at 190 cm close to the end of the Last Glacial Maximum (LGM);
6. minor IRD peak during the LGM;
7. IRD peak at 290 cm (LGM);
8. IRD peak at 350 cm (LGM);
9. IRD peak at 440 cm (LGM);
10. IRD peak at 500 cm associated with meltwater discharge (LGM);
11. moderate IRD peak at 530 cm associated with meltwater discharge (LGM).

Samples 5–11 illustrate in very high detail the composition, and compositional changes, of IRD during the LGM. Massive amounts of icebergs drifted from the Fennoscandian Ice Sheet into the Norwegian Sea during (1) very cold periods, (2) meltwater peaks, and (3) major deglaciations.

The following discussion of the IRD composition aims to resolve two major questions:

(1) can we identify specific sub-regions underneath the FIS where the majority of the icebergs came from?; and
(2) can we determine which surface circulation changes occurred during the Late Weichselian Glaciation?

The identification method is stepwise elimination of unlikely sources by comparing the known bedrock geology of Norway (Oftedahl, 1980; Rokoengen et al., 1980; Sigmond et al., 1984) to the samples' IRD composition, thereby reducing the number of possible candidates to the most likely ones. This method requires a detailed sample-by-sample discussion of the IRD composition, which may make some text passages difficult to read. But the task of linking IRD assemblages to their continental origins without the help of actual source area samples (Darby and Bischof, 1996) is challenging, to say the least. Therefore, the following passages are not exactly 'light reading'. The major sources of icebergs from the FIS changed between the Precambrian granite and gneiss terrains of southwestern Norway and adjacent parts of Sweden and Denmark, the Caledonian metasediments of western Norway, the metapelite formations of central western Norway, and the intrusive rocks of northern Norway and Finnmark.

Table 5.7 shows the composition of all 11 samples and all 64 distinguished IRD lithologies. Crystalline rock fragments, quartz and feldspar grains are most abundant. The crystalline rocks account for 38–98% of all rock fragments. With the exception of two samples that were heavily influenced by icebergs with clastic sedimentary rocks from the Barents shelf and Svalbard, the crystalline rock proportion in 9 out of the 11 analysed samples varies between 79 and 98% of the rock fragments. This places the origin of most of the IRD, and therefore the icebergs, into the crystalline formations of western Scandinavia and the northwestern British Isles. The crystalline rocks are mostly metamorphic rocks (34.5–82.5% of the rock fragments), primarily gneisses (26–75%), quartzites (0.7–19.6%), phyllites, and mica

Table 5.7.

Core number	23074	23074	23074	23074	23074	23074	23074	23074	23074	23074	23074
Sample number**	1	2	3	4	5	6	7	8	9	10	11
upper boundary (cm)	1	10	20	118	210	260	310	360	460	510	545
lower boundary (cm)	9	18	35	140	230	280	330	380	480	530	565
depth at midpoint	5	14	27.5	129	220	270	320	370	470	520	555
Age, top*	9500	11000	11836	16808	19088	19825	20562	21299	23050	24003	25033
Age, bottom	10600	11645	13741	17943	19383	20120	20857	21594	23431	24577	25641
age at midpoint	10050	11323	12789	17376	19235	19972	20710	21447	23240	24290	25337
*in calendar years before the present (Voelker, 1999; Vogelsang, 1990).											
light grey sandstones	1.86	1.63	2.28	1.73	0.2	0.6	0.22	0.27	0.19	1.02	
grey siltstones	2.79	2.34	22.83	21.61	4.29	4.77	2.69	2.17	2.61	1.54	0.18
black shales	1.27	1.09	1.83	2.69	2.24	0.6	0.45	0.27	0.75	0.85	0.18
green siltstones	0.42	0.22		0.1							
brown siltstones	1.18	0.71	1.37			0.2	0.22		0.37	0.34	
red siltstones	1.78	1.74	0.91			0.4	0.67		0.56	2.05	0.36
brown clay-ironstones				0.19			0.67	0.27			
cherts	0.25			0.38	0.2		0.67		0.19		0.18
flint (Feuersteine)		0.27		0.96		0.4		0.27			
chalk		0.27			0.61	1.39	0.67	0.27	1.12	0.17	
limestones/dolomites		0.27	2.74	0.67	0.41	0.6	1.35	1.36	0.37	2.39	
granites w/ white feldspar	0.08							0.54			
granites, white fsp-qtz	0.76	0.60	5.94	0.86	0.61	0.6					1.25
granites w/ red feldspar		0.05				0.2					
granites, red fsp-qtz	0.42	0.60	4.11	0.48	0.2	0.2	0.22			0.17	0.54
granites w/ yellow feldspar		0.05									
granites w/ yellow fsp-qtz	0.42	0.49	0.46							0.34	0.18
granites, white/red feldspar	0.08	0.11		0.29							
granites, white/red felds-qtz	0.93	0.27		0.1	0.41					0.17	0.18
granites, white/ yellow fsp		0.05									
granites, white/ yellow fsp-qtz	0.17					0.2					
diorites					0.82	0.6		0.27		0.34	0.71
dior/gabbros >50% mafics	0.17	0.16							0.19	0.68	0.36
gabbros >90% mafics	0.08	0.33	0.91			0.99	0.22			1.37	0.71
pyroxenites		0.11									
grey phyllites	0.51	0.33	0.91	0.29	0.61	0.99	0.67	1.36	1.12	0.85	

Core number	23074	23074	23074	23074	23074	23074	23074	23074	23074	23074	23074
Sample number**	1	2	3	4	5	6	7	8	9	10	11
upper boundary (cm)	1	10	20	118	210	260	310	360	460	510	545
lower boundary (cm)	9	18	35	140	230	280	330	380	480	530	565
depth at midpoint	5	14	27.5	129	220	270	320	370	470	520	555
Age, top*	9500	11000	11836	16808	19088	19825	20562	21299	23050	24003	25033
Age, bottom	10600	11645	13741	17943	19383	20120	20857	21594	23431	24577	25641
age at midpoint	10050	11323	12789	17376	19235	19972	20710	21447	23240	24290	25337

*in calendar years before the present (Voelker, 1999; Vogelsang, 1990).

mica schists, silver	0.51	0.05	0.46	0.19	0.2	1.99	1.12	1.08	0.56		
green schists					0.41	0.6	0.67		0.19		
serpentinites	0.08	0.11	0.46								
white quartzites	2.70	4.41	3.65		1.02	1.99	3.14		1.31	1.19	
grey quartzites	0.76	1.14	2.28	0.1	0.61	1.19	0.9				
black quartzites	0.25	0.33	0.91								
green quartzites	1.10	0.71	0.46	0.19		0.4					
brown quartzites	0.08	0.33									
red quartzites	1.69	1.85	0.91	0.1			0.22	0.27		1.02	0.18
orange quartzites		0.16		0.19						0.34	
yellow quartzites		0.44									
leucocratic gneiss	18.51	17.20	12.79	11.34	17.96	18.29	20.4	17.89	15.49	18.43	23.04
mafic gneiss	4.90	5.61	6.85	3.75	7.76	10.14	12.33	11.38	15.67	12.63	8.93
white pumice	0.08							0.27			
black volc. glass		0.27									
grey/green volcanic glass	0.08		0.46								
brown volc. glass						0.6					
grey basalt	1.52	0.33				0.2	0.22	0.27	0.19	0.34	0.18
black basalt	1.86	0.71	1.37			0.8		0.54	0.19	2.56	2.68
dark green volcanics				0.1		0.2				0.51	6.79
brown volcanics	0.51	0.38									
red volcanics	0.42	0.65				0.2	0.22	0.27			
angular quartz	5.07	7.19	3.65	8.26	9.39	4.57	5.83	7.86	5.41	4.1	11.61
rounded quartz	23.16	17.91	14.16	32.37	32.45	33	29.37	33.33	36.94	26.45	25.54
round/oval quartz	1.69	3.21	1.83	2.4	3.88	1.59	2.02	2.98	3.17	4.78	
white felds., angular	1.27	1.42		0.29	0.61	0.4	0.45		0.37	0.85	0.54
white felds., rounded	12.17	11.21	4.57	5	10.41	9.54	9.42	12.2	10.63	9.73	10.54
white feldspar, round	0.76	0.44		0.29	0.2	0.2	0.45		0.19	0.17	
red/orange felds/ang.		0.54					0.9		0.37	0.17	0.18
red/orange felds/rded	4.06	6.37	0.91	3.36	1.84	0.8	1.79	2.17	0.75	1.88	3.57
red/orange felds/rd	0.17	0.16								0.17	0.18
yellow felds/angular	0.17	0.11				0.2	0.22				
yellow feldsp/rounded	2.20	3.43		0.96	0.82	0.4	0.45	1.08	0.19	0.51	0.71
yellow felds. round	0.25	0.27								0.34	
black minerals				0.1	0.2					1.19	0.36
garnet, deep red				0.1						0.17	
olivine, bottle-green									0.37		0.18
coal fragments	0.76	1.36		0.48	0.2		0.9	1.36	0.56		

**as referred to in the text

Core number	23074	23074	23074	23074	23074	23074	23074	23074	23074	23074	23074
Sample number**	1	2	3	4	5	6	7	8	9	10	11
upper boundary (cm)	1	10	20	118	210	260	310	360	460	510	545
lower boundary (cm)	9	18	35	140	230	280	330	380	480	530	565
depth at midpoint	5	14	27.5	129	220	270	320	370	470	520	555
Age, top*	9500	11000	11836	16808	19088	19825	20562	21299	23050	24003	25033
Age, bottom	10600	11645	13741	17943	19383	20120	20857	21594	23431	24577	25641
age at midpoint	10050	11323	12789	17376	19235	19972	20710	21447	23240	24290	25337
*in calendar years before the present (Voelker, 1999; Vogelsang, 1990).											
Lithologic diversity	45	51	27	32	27	36	32	25	28	35	27
Diversity of sedimentary rocks	8	10	6	9	7	8	10	8	9	7	4
Diversity of crystalline rocks	26	29	16	13	11	19	12	11	9	15	13
Diversity of mineral fragments	11	12	5	10	9	9	10	6	10	13	10
Percent of all rock fragments											
limestones and dolomites	0.00	0.57	3.66	1.43	1.06	1.22	2.76	3.37	0.89	4.85	0.00
chalk and flint	0.00	1.14	0.00	2.05	1.57	3.63	1.37	1.34	2.69	0.34	0.00
other cherts	0.52	0.00	0.00	0.81	0.52	0.00	1.37	0.00	0.46	0.00	0.39
plutonic rocks	6.38	5.93	15.24	3.70	5.26	5.65	0.90	2.01	0.46	6.23	8.43
volcanic rocks	8.97	4.90	2.44	0.21	0.00	4.05	0.90	2.67	0.91	6.92	20.69
gneisses	47.76	47.78	26.22	32.25	66.36	57.62	67.01	72.49	74.85	63.00	68.56
quartzites	13.45	19.61	10.98	1.24	4.21	7.26	8.72	0.67	3.15	5.17	0.39
phyllites and mica schists	2.07	0.80	1.83	1.03	3.15	7.26	5.04	6.04	4.49	1.72	0.00
all crystalline rocks	78.97	79.25	57.32	38.43	78.97	81.84	82.58	84.55	83.86	83.04	98.07
all sedimentary rocks	21.03	20.75	42.68	61.57	21.03	18.16	17.42	15.45	16.14	16.96	1.93
clastic sedimentary rocks	18.97	16.19	39.02	56.25	17.36	13.32	10.07	7.38	10.76	11.76	1.54
metamorphic rocks	63.45	68.42	39.63	34.52	73.71	72.13	80.77	79.20	82.49	69.90	68.95
igneous rocks	15.34	10.83	17.68	3.91	5.26	9.71	1.80	4.68	1.37	13.14	29.12
Percentages of mineral fragments											
quartz	58.71	54.17	78.18	80.99	76.45	77.24	73.12	74.09	77.96	69.95	69.56
white feldspar	27.86	25.00	18.18	10.50	18.76	20.00	20.28	20.46	19.16	21.28	20.75
reddish feldspar	8.29	13.54	3.64	6.32	3.08	1.58	5.28	3.64	1.92	4.40	7.36
yellow feldspar	5.14	7.29	0.00	1.81	1.37	1.18	1.32	1.81	0.33	1.68	1.33
black minerals	0.00	0.00	0.00	0.19	0.33	0.00	0.00	0.00	0.00	2.36	0.67
garnet	0.00	0.00	0.00	0.19	0.00	0.00	0.00	0.00	0.00	0.34	0.00
olivine	0.00	0.00	0.00	0.00	0.00	0.00	0.00	0.00	0.63	0.00	0.34
feldspar	41.29	45.83	21.82	18.63	23.21	22.76	26.88	25.91	21.41	27.36	29.43

IRD composition of all counted lithologies and minerals, in percent of the >500 μm fraction, of kastenlot core 23074 from close to the Norwegian continental margin. Note that volcanic rocks are almost missing in samples 4 and 5, which otherwise are characterised by the highest clastic sedimentary rock percentages in the entire core.

schists (0–7.3%). Plutonic rocks, including both leucocratic (granites, tonalites, syenites, granodiorites, etc.) and mafic (diorites, gabbros, norites, ultramafics) types, constitute 0.9–15.3% of the rock fragments. Sedimentary rocks are very abundant (43–62% of the rock fragments) between 18 and 11.8 ka (samples 3 and 4) when the Barents Ice Sheet disintegrated, but are relatively low (2–21%) in all other samples. The lowest value of 1.9% is from the oldest sample 11 which was deposited between 25.6 and 25 ka. The sample consists of remarkably pure crystalline material and is devoid of carbonates, chalk, flint and metapelites. This effectively removes southern Scandinavia (primarily Sweden and Denmark) where carbonates are common, and most of northern Norway where metapelites occur, from the list of potential sources. Furthermore, the sample is low diverse, with only 27 of 64 lithologies present, and it contains the highest proportion of volcanic rocks in this core, 20.7% of the rock fragments. The volcanic rocks have a fresh appearance and, for reasons explained,[19] are coming from either the Faeroer Islands or from some submarine outcrop on the bottom of the southern Norwegian shelf. Taken together, these characteristics narrow the source(s) of sample 11 down to a few, geographically limited places in southwestern Norway and the northwestern British Isles, which were both glaciated at this time. The icebergs drifted straight towards the north.

But with the exception of this sample and the two samples that were influenced by the Barents Ice Sheet decay, the proportion of sedimentary rocks fluctuates between 15.5 and 21% of the rock fragments. That means that during most of the last 15,000 years of the Weichselian Glaciation, a considerable portion of the IRD, and for that reason a considerable number of icebergs, came from sources outside the crystalline terrain of continental Norway. Potential candidates are the outer Norwegian shelf, the northern British Isles, the southwestern Baltic region, and Svalbard and the Barents shelf. Between one-sixth to one-fifth of the icebergs came from some of these sources.

Chalk and flint, indicator rocks of a southern provenance, are present in eight of the samples but are missing in samples 11, 3 and 2. Therefore, during most of the LGM, icebergs from the southern margin of the Fennoscandian Ice Sheet occasionally drifted northwards into the eastern Norwegian Sea. A maximum southeastern influence (factor 6 in Table 5.5) is recorded in sample 6 (chalk and flint: 1.8% of the total IRD and 3.6% of the rock fragments) for the time interval of 20100–19830 calendar years bp. The sample contains largely gneisses, quartz and white feldspar. Clastic sedimentary rocks are rare and dominated by grey siltstones. Light coloured intrusive rocks, usually a common occurrence in Weichselian IRD of the Norwegian Sea, occur only sporadically. On the other hand, this sample has the second highest gabbro concentration of this core (only sample 10 has more), and there are relatively high amounts of silver, grey and green metapelites. This points to an origin of at least a part of this sample's IRD in the metapelite formations of western Norway between 65 and 68 degrees North, but there may be other possibilities. Among quartzites, only white, grey and green varieties were found. This is unusual, because most Norwegian Sea IRD samples contain yellow, orange, red, and brown quartzites (Figs 5.7a, b), and the ratios of green to yellow, orange, red and brown quartzites are for the most part lower than 2. In fact, of the 46 samples shown in Fig. 5.7b, 34, or 74%, have ratios of less than 1, that is, they contain more

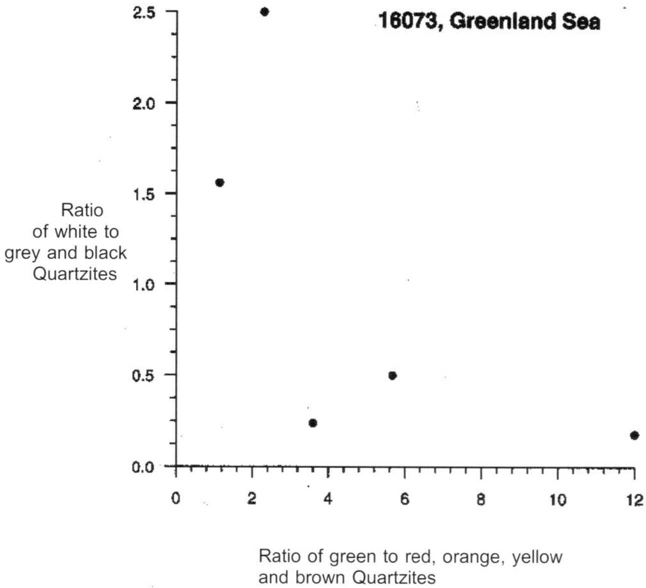

Fig. 5.18. Composition of quartzites, ranked by colour, in core 16073, Greenland Sea. Note the three samples at the bottom right which are characterised by high amounts of grey, green and black quartzites, but comparatively low amounts of white, yellow, orange, red and brown quartzites. These three samples are of Holocene age and were thus deposited by ice that drifted with the cold, sea ice-covered East Greenland Current. Much, if not all, of this ice ultimately originated in the Arctic Ocean, so that the quartzite colour distribution shown here represents Arctic-derived IRD. Compare this to the colour distribution of Norwegian Sea quartzites as shown in Fig. 5.7a, and it becomes clear that the Norwegian Sea quartzites contain much more of the yellow, orange, brown and red types. Thus, the colour distribution of the quartzites in sample 9 of core 23074 resembles more closely Arctic-derived IRD than IRD from the Norwegian side. This claim is further supported by Fig. 5.7. In this figure, the only data points to the lower right belong to samples from oxygen isotope stages 6 and 8 which are high in clastic sedimentary rocks and coal. For these samples, an Arctic origin was determined on the basis of the vitrinite reflectance ratios of the coal fragments (Bischof et al., 1990).

yellow, orange, brown and red quartzites than green ones. Sample 6 would be at the bottom to the far right, far removed from the majority of the Norwegian Sea IRD samples. But IRD in the Greenland Sea contains relatively more green, grey and black quartzites than yellow, orange, brown and red types, producing ratios to the lower right as those in sample 6 of core 23074 (Fig. 5.18). The data points farthest to the lower right are of Holocene age and were deposited by ice in the cold East Greenland Current which carries ice from the Arctic Ocean. For this reason, sample 6 of core 23074 may contain a small but recognizeable fraction of IRD from the Arctic Ocean.

Figure 5.19 shows the concentration of the most important IRD components of core 23074 plotted against the new, revised time-scale of Voelker (1999) and

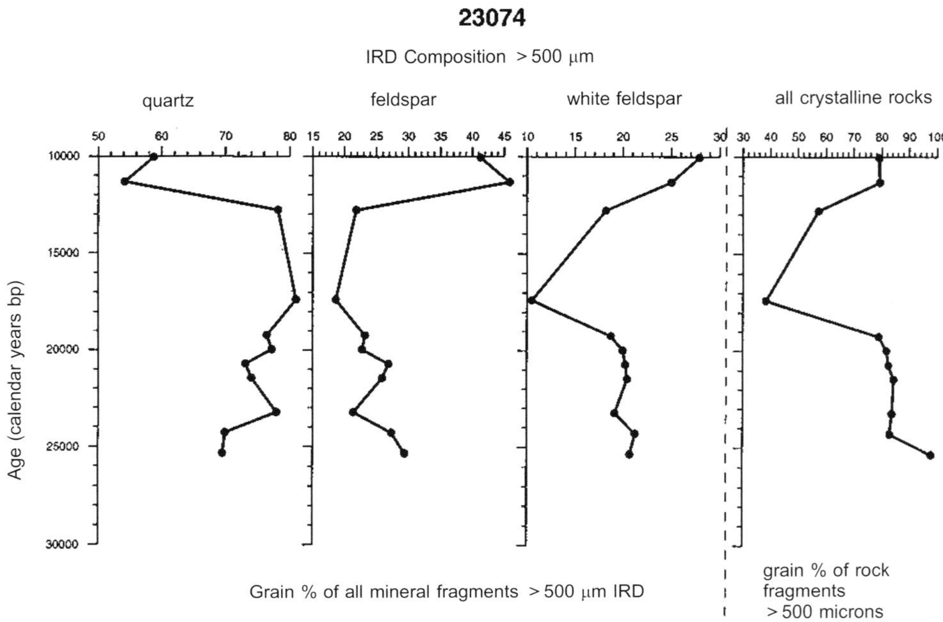

Fig 5.19. Composition of major IRD lithologies and mineralogies in core 23074, off the Mid-Norwegian continental margin.

(a) Concentration of crystalline rocks among rock fragments >500 μm (excluding quartz and feldspar, right side), and composition of mineral fragments >500 μm (in percent of the minerals, excluding rock fragments). Among the mineral fragments, quartz is clearly most abundant with 70–80% in sediments older than 12,000 years bp. At that time, the quartz proportion among mineral fragments dramatically dropped, and this drop is compensated by rising amounts of feldspar. The quartz and feldspar curves are exact mirror images of each other: if the one component drops, the other one rises and vice versa. Other minerals such as garnet, olivine, pyroxenes and amphibole play an unimportant role. White feldspar and crystalline rocks show the same percentage trends. Initially they both stay at high and constant levels before they drop between 19 and 18 ka. Thereafter, the values of crystalline rocks rise back to the level before 19 ka, whereas the white feldspar proportion rises to a significantly higher level. Interestingly, the dramatic drop in both crystalline rocks and white feldspar fragments is not reflected by a similarly drastic change in the proportion of quartz to feldspar. Between these mineral components the most drastic change occurs after 13 ka, close to the end of the Weichselian Glaciation.

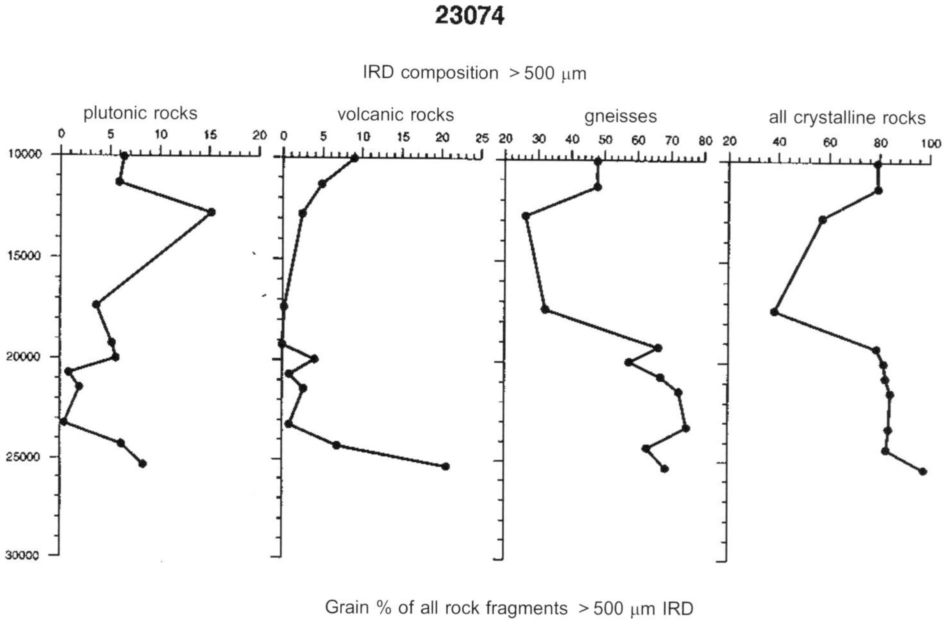

(b) Composition of crystalline rock IRD > 500 μm (in grain % of rock fragments). The three graphs on the left for plutonic and volcanic rocks and gneisses show how heterogenous the crystalline rocks are. Plutonic rocks can be as low as 0.5% in sample 9 (23.4–23 ka), or as abundant as 15.2% in sample 3, which represents the last phase of the Younger Dryas cold period between 13,740 and 11,830 years bp. Even without knowing the exact source of these plutonic rocks, it is clear that the relative importance of this source was 30-times stronger during the Younger Dryas than between 23 and 23.4 ka. At the earliest, at 26 ka, volcanic rocks account for 20% of the crystalline rocks. These volcanics are grey, black and dark green, fresh-looking basalts of most probably Tertiary age. For reasons explained in,[18] I find an origin from the Faeroer Islands or from submarine outcrops on the Norwegian shelf off Kristiansund (Bygge et al., 1980) most likely.

Vogelsang (1990). For reasons explained in endnote 18, the volcanic rocks at the bottom of the analysed section (26–24 ka) came from a Tertiary volcanic terrain to the southeast of the Norwegian Sea, most likely the Faeroer Islands or a submarine outcrop on the Norwegian shelf. Therefore, at this time the ice sheets in southern Norway – and perhaps also the northwestern British Isles – must have been in retreat, producing icebergs and meltwater.

The icebergs drifted (a) in a current that moved northward fast enough to reach the Vøring-Plateau before significant melting of the icebergs could occur, and (b) in a watermass that was cold enough to prevent them from melting. It is my proposition that the water temperature fluctuated between 2 and 6° C between 26 and 24 ka, and that the icebergs drifted in an ocean that was ice covered in the winter but opened up during the summer. The scenario was probably similar to the Baffin Bay of today. Cold, sea ice-covered winters would permit the northward passage of the icebergs,

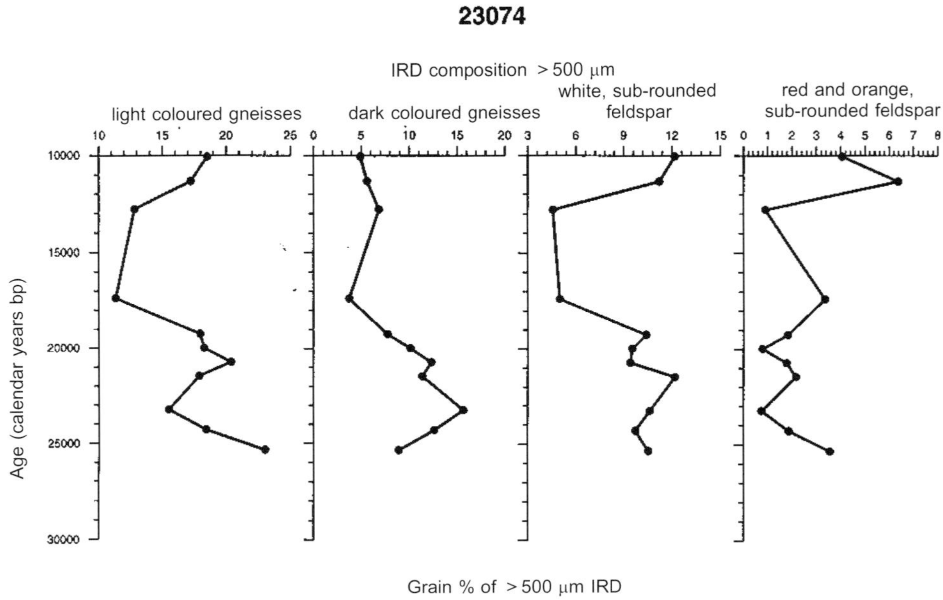

(c) Composition of crystalline IRD components in grain % of all > 500 μm IRD. From left to right: light coloured gneisses, dark coloured gneisses, white feldspar, and red and orange feldspars. The feldspars are listed as subrounded grains here. It is interesting to note that at 18 ka, when the proportions of most crystalline IRD lithologies drop in response to the arrival of icebergs from the Barents Ice Sheet, the amount of reddish-coloured feldspars rises. Considerable percentage changes of the crystalline components depicted here occurred between 25 and 19 ka, a time when the overall percentage of crystalline rocks among the rock fragments > 500 μm remained the same (b, right).

and melting and open water during the summer would facilitate the release of particles from the icebergs.

The IRD of sample 11 is almost entirely composed of crystalline rocks (98% of the rock fragments, Table 5.7 and Fig. 5.19b), of which 69% are metamorphic rocks. Therefore, the icebergs came almost exclusively from the metamorphic terrains of Norway and the northwestern British Isles. The complete lack of carbonates, chalk, flint and metapelites shows that the icebergs did not come from north of 65° of Norway or from the Baltic region. The low diversity of the IRD in sample 11, with only 27 out of a total of 64 distinguished IRD petrographies present (Table 5.7), further underscores the limited extent of glacial iceberg sources at this time. Only a small segment of the Fennoscandian Ice Sheet actively produced icebergs which drifted to the site of core 23074. At the same time, other parts of the same ice sheet were either stagnant, not producing any icebergs, or icebergs from other parts of the Fennoscandian Ice Sheet drifted around the site of 23074, not affecting the IRD composition there. A distinction between these two possibilities will be made later. This constrains the sources of the icebergs to valley glaciers coming out of the fjords

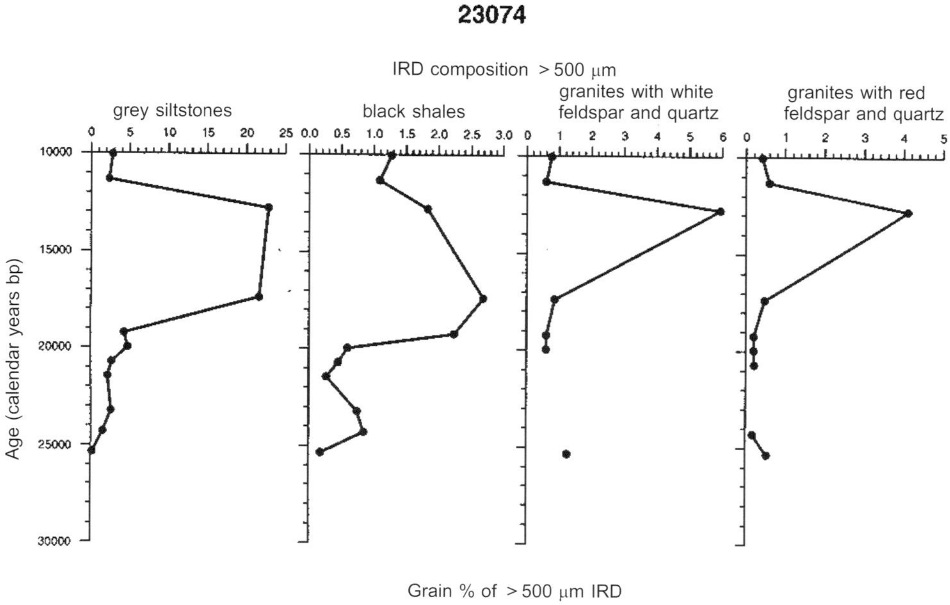

(d) Composition of grey siltstones, black shales, granites composed of white feldspar, and granites with red feldspar against time. Both types of clastic rocks are low before 20 ka, after which both components experience a 4–5 fold increase. Notably, the rise of the black shales precedes the increase of grey siltstones by 1–2000 years. White and red granites, the most abundant rock types among light coloured plutonic rocks in Norwegian Sea IRD, show a completely different occurrence pattern. Both are low until 17 ka, but then sharply rise in abundance at 13 ka, during the later part of the Barents Ice Sheet decay. This means that, wherever the exact source of these granites may be, its relative influence on the IRD composition increased by 600–800% between 17 and 13 ka. This particular type of IRD composition with high granite proportions was isolated as an independent factor by factor analysis. It is factor 3 in Table 5.5. The highest score of this factor on any core sample was obtained for sample 16052, 21–26 cm. This sample was deposited between 16.4 and 14.5 ka and contains 68.3% clastic sedimentary rocks. It is undeniably from the Barents shelf and adjacent areas, and factor 3 therefore represents plutonic rocks of northern Norway or from Svalbard.

of southwestern Norway and northwestern Britain. The drift speed (or current velocity) necessary to carry the icebergs north between two consecutive melting seasons is 5.2 cm/sec, almost identical to the present-day drift speed of the Arctic sea ice cover (Rigor and Ortmeyer, 2000; assumptions: distance from the NW British Isles to the site of 23074 = 1200 km; time between two consecutive melting seasons = 9 months).

Figure 5.19b shows how strongly the influences from different sources composed of crystalline rocks varied through time. The proportion of plutonic rocks in sample 3 is 30 times higher than the one in sample 9. Thus, the influence from the source area that contains the plutonic rocks must have increased by a factor of 30 between

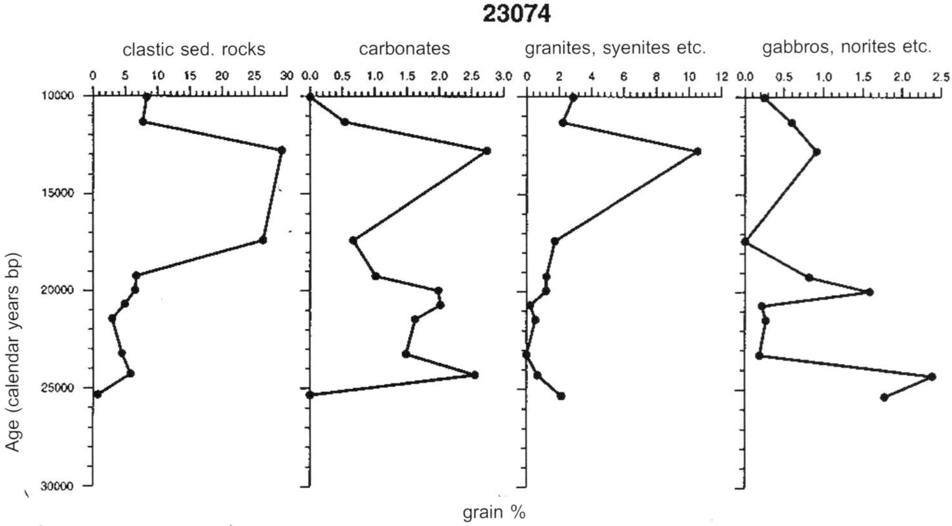

(e) Clastic sedimentary rocks, carbonates, light and dark coloured plutonic rocks in >500 µm IRD versus age in calendar years bp. The most important feature is that clastic rocks account for less than 7% of the >500 µm IRD from 25 ka until 19 ka. During the same time, the Fennoscandian Ice Sheet advanced onto the Norwegian shelf (Baumann et al., 1995). If converted into grain % of the rock fragments >500 µm, the clastic sedimentary rocks make up 7–17% of the IRD (Table 5.7). This sets limits to the amount of clastic sedimentary rocks resulting from glacial erosion of the outer Norwegian shelf, circa 7.4% of the total >500 µm fraction, and up to 17% of the rock fragments *sensu stricto*. Immediately after 18 ka, the values rise steeply owing to the arrival of icebergs with petrographically different clastic sedimentary rocks from the disintegrating Barents Ice Sheet.

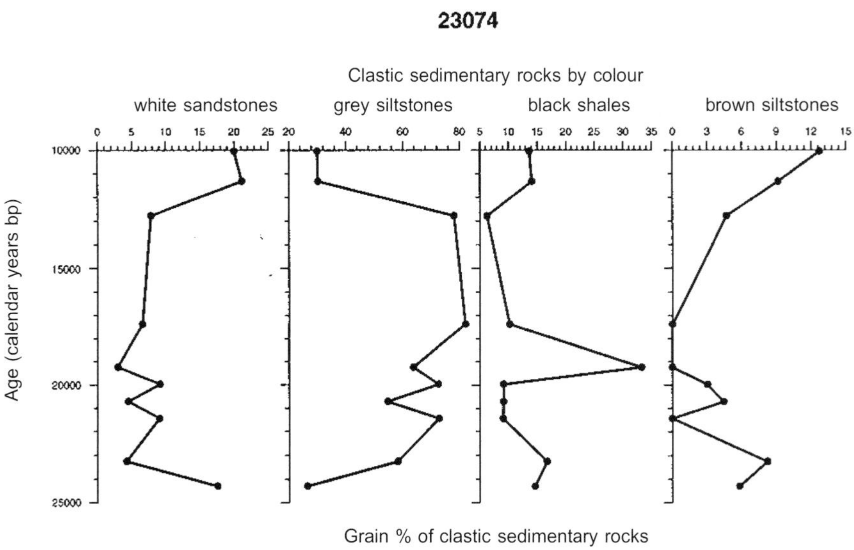

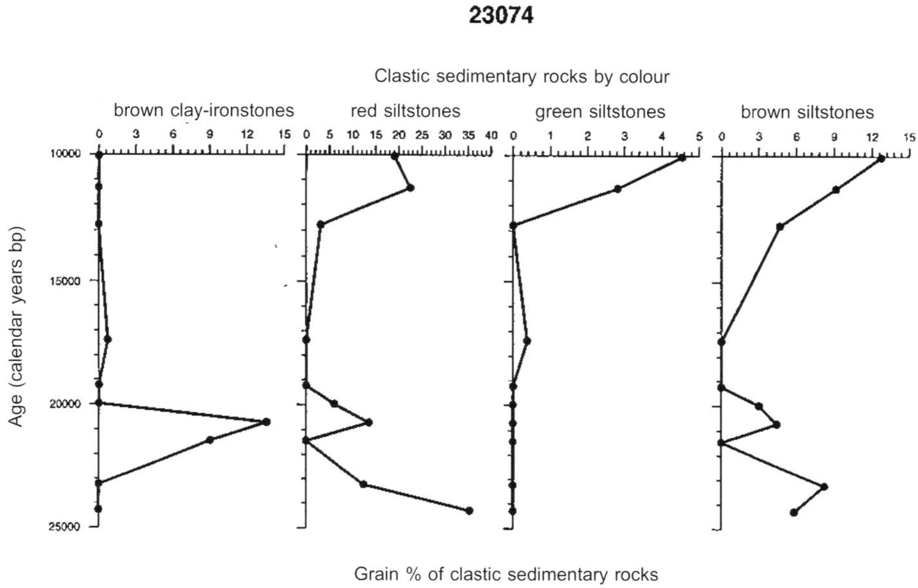

(f and g) Composition of clastic sedimentary rocks in grain % of the clastic sedimentary rocks. Grey siltstones dominate during the time of the Barents Ice Sheet decay from 18 to 12 ka, when they account for 80% and more of the clastics. The proportion of grey siltstones rose gradually from 25% at 25 ka to 80%, showing that there is not only a rise in the overall proportion of the clastic rocks among IRD, but also a temporal evolution from various multicoloured clastics to a comparatively monochromatic assemblage in which grey siltstones dominate. The Barents Ice Sheet decay is preceded by an unusually high amount of black shales.

23 ka and 13.7 ka. Factor analysis has isolated factor 3 (Table 5.5), which must represent a plutonic terrain somewhere in the vicinity of the Barents shelf, either Svalbard or northern Norway. In either case it is a northern source. Sample 3 in core 23074 contains very small amounts of glacial IRD, which is low diverse (only 27 of 64 distinguished lithologies present) and contains no chalk or flint. The sample therefore has no contribution from the Baltic region. On the other hand, the sample contains 43% sedimentary rocks (in percent of the rock fragments), including 3.7% carbonates. This means that at this time at least 43% of the IRD (and the icebergs) came from Svalbard and adjoining parts of the Barents shelf. But the parent ice mass that produced these icebergs must also have overridden some of the crystalline rocks which make up approximately 50% of the surface of the Svalbard archipelago. Therefore the true proportion of icebergs from Svalbard towards the end of the Younger Dryas was probably higher than 43%, probably in the order of 70–80%. The Younger Dryas cold period was a brief return to semi-glacial conditions during which iceberg rafting was limited. The few icebergs came from Svalbard and drifted southward through the Norwegian Sea. From the first recorded arrival of large quantities of icebergs with clastic sedimentary rocks from the Barents shelf at 18 ka until the end of the Younger Dryas, at 11.7 ka, the Norwegian Sea was under the

influence of a surface circulation that carried icebergs from north to south. This system existed for 6300 years.

5.5.2 Different crystalline source terrains

The crystalline rocks that were dropped by icebergs between 25 and 10 ka were not a homogenous assemblage whose composition never changed. In fact, the composition of the crystalline rocks changed as much as their relative proportion among the > 500 μm IRD. Figure 5.19b shows that the crystalline rocks stay at a conservative level of around 80% (of the rock fragments) for most of the time, except for the lowest sample which is composed almost entirely of crystalline rocks from southwestern Norway and the northern British Isles, and samples 3 and 4 which were affected by the Barents Ice Sheet decay. While the abundance curves of gneisses and white feldspar (Fig. 5.19c) look similar to the curve of all crystalline rocks, the curves for plutonic and volcanic rocks show entirely different patterns (Fig. 5.19b), indicating that the icebergs from the Fennoscandian Ice Sheet came from different sub-sections of the Scandinavian Shield and the Norwegian Caledonides at different times. This becomes clear from the relative amount of light coloured gneisses to dark ones (Fig. 5.19). The lowest four samples of the core, samples 8 through 11, show percentage changes opposite to each other; as the proportion of mafic gneisses rises from 25 to 23 ka, the relative amount of light coloured gneisses drops. Thereafter the trend reverses and decreasing amounts of dark gneisses are balanced by increasing amounts of light gneisses. From that point on until 13 ka, the percentage fluctuations are more or less the same before the trend reverses again. The anti-phase relationship at the bottom as well as near the top can only mean that the iceberg sources gradually moved from a terrain in which light coloured gneisses are more abundant than dark ones to another terrain where the proportion of dark gneisses is higher. To my knowledge, literature in which the Norwegian bedrock geology is described in such detail (that is, accurately listing the quantitative proportion of light and dark coloured rocks among the Norwegian gneisses) to make it possible to pinpoint the source, is not available. Therefore, instead of being able to point to the source, one must simply be content to state that a shift from one part of the Norwegian gneiss terrains to another, and back, occurred between 25 and 21.5 ka. For the period from 21.5 until 13 ka, the percentage changes of both dark and light coloured gneisses have the same sign, which means in this case that they were derived from the same general area.

White and red granites are rare from 26 to 17 ka (Fig. 5.19d). Between 24 and 21 ka, none were found at all. But within the time period from 17 to 12 ka, their proportion rose from 1.5 to 10%, by a factor of 7. Directly after that peak, they drop again to values similar to the ones at the bottom. These granites are only abundant during the late stage of the Barents Ice Sheet decay when massive quantities of icebergs drifted to the south. Therefore, they most probably came from a northern source in northern Norway or Svalbard (factor 3 in Table 5.5).

Mafic intrusive rocks do not show the same abundance pattern as their light coloured counterparts (Fig. 5.19e). Rather, they show two abundance peaks at 24 and 20 ka, times when the proportions of granites, syenites, etc. were very low. Therefore the gabbros and norites came from a different source area than the one represented by factor 3 (Table 5.5). Most likely this was factor 4 which has high loadings on mafic intrusives and gneisses. And gneisses are very abundant during the two times of high percentages of mafic intrusives. However, the factor scores of individual samples do not allow the source to be pinpointed.

Quartz and feldspar are the two most important mineral grain varieties. In core 23074, quartz is by far more abundant than feldspar. From 26 to 12 ka, 70–80% of the mineral grains are quartz with the remainder being feldspar (Fig. 5.19a). But between 13 and 12 ka, a dramatic change occurs to approximately equal proportions of the two mineral types, and this occurred directly after the light coloured plutonic rocks rose to their respective peaks (Fig. 5.19d).

The fact that four different IRD lithologies peaked in sequence shows the changing ice drift directions and iceberg sources. Until 19 ka, crystalline rocks, primarily gneisses from the southeastern margin of the Norwegian Sea, were dominant. Between 19 and 17 ka, clastic sedimentary rocks from the Barents shelf rose to peak values. While the clastic sedimentary rocks stay high until 12 ka, white and red granites peak between 13 and 12 ka. Thereafter, clastic and light coloured intrusive rocks drop (Fig. 5.19e) as feldspars rise (Figs. 5.19a, c). The origin of the feldspar-rich IRD can be extracted directly from the geographic distribution of feldspars in our samples. The highest feldspar content (39.5%) and the highest proportion of feldspar among mineral fragments (61.5%) were found in giant box core 23075 at a depth of 20–39 cm. This core is located at the mid-Norwegian continental margin at 66°21 North and 6°08 East in a water depth of 652 m, close to the margin of the Fennoscandian Ice Sheet at the time the sample was deposited, between 13 and 11 ka. The sample thus represents debris from coastal Norway and parts of the shelf north of 65°. The most striking characteristic of this IRD is that it contains more feldspar than quartz. Samples in which feldspar accounts for more than 30% of the mineral fraction cannot be derived from anywhere else than continental (western) Norway, for the following reasons: IRD in the Greenland Sea contains much more quartz than feldspar (Fig. 5.20). There is no overlap between the data point distribution of IRD in the Greenland Sea and IRD in the Norwegian Sea from the time period 25–15 ka. Furthermore, IRD in shallow marine sediments from close proximity of Svalbard, Northeast Greenland and Franz Josef Land generally contained less than 7% feldspar among the mineral grains. Only one sample from north of Spitsbergen had 27% feldspar among the minerals (endnote 15).

Therefore the ice drift directions changed as follows: first, before 19 ka, icebergs came mainly from southwestern Norway, the Baltic region and parts of the northern British Isles. Within this general area, the iceberg sources have changed. At times, the icebergs came more from an area where the bedrock contains relatively more dark coloured gneisses. At other times, the iceberg sources shifted to an area with more light coloured gneisses. Even though it is not possible to determine the location of either area, the shift of the iceberg sources itself remains beyond dispute. After 19 ka, the iceberg sources moved to a region with abundant clastic sedimentary rocks, the bottom of the Barents Sea and parts of Svalbard. The southward drift of icebergs from the Barents shelf to the site of 23074 lasted approximately 6000 years. Towards its end, light coloured plutonic rocks from northern Norway were transported along

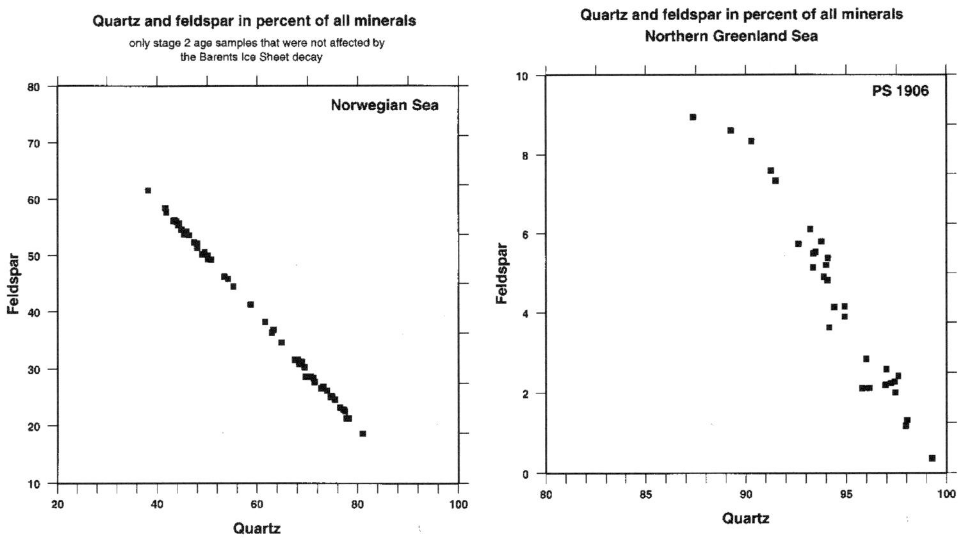

Fig. 5.20(a). Composition of the mineral fraction of Norwegian Sea IRD for the time period from approximately 25 ka to 15 ka. Only samples with at least 100 quartz and feldspar grains are shown. The purpose of this plot is to show that the Norwegian Sea crystalline IRD is truly exclusively derived from Norway and not somewhere else. The effect of the Barents Ice Sheet was filtered out by removing all samples with over 25% clastic sedimentary rocks. All data points are on a straight line because the minerals consist of quartz and feldspar and nothing else. Furthermore, feldspar constitutes at least 18% and up to almost 62% of the mineral fraction in the Norwegian Sea.
By contrast, Greenland Sea IRD from core PS 1906 (76°50 N and 2° W; Fig. 5.20b) is dominated by quartz, which accounts for at least 87% of the minerals, and the data points are no longer on a straight line but scatter somewhat because of minerals other than quartz and feldspar (pyroxenes). If both data sets are plotted on the same scale, there is absolutely no overlap. They are distinct, different populations.

with icebergs from the shrinking Barents/Svalbard ice complex on a more southwestward ice drift, rather than straight south as before. At 12 ka, another very pronounced source shift occurred: the supply of icebergs with clastic rocks from the Barents Ice Sheet subsided, and the icebergs originated from glaciers in central western Norway, drifting westwards.

5.5.3 Sedimentary rocks from the Norwegian shelf

After 25 ka, the glacier cover over northwestern Europe grew in geographic extent and glaciers began to override and erode the sedimentary bedrock which is exposed on the bottom of the Norwegian shelf. Therefore, the icebergs that broke off the outer ice margin at this time should contain sedimentary rocks in proportions higher than before. This shows in the rising proportions of both clastic sedimentary rocks and carbonates among IRD between 25 ka (sample 11) and 24 ka (Fig. 5.19e, Table 5.7). Clastic sedimentary rock percentages were practically negligible before 25,000 years bp, but during the time of maximum ice extent in Scandinavia when ice reached the outer shelf (between 26 and 20 ka, Baumann et al., 1995), the clastic sedimentary rock percentages increased. While their relative proportion among all > 500 μm IRD stays at values of mostly less than 5%, their proportion among rock fragments varies between 7.4 and 17.4% from 24.6 ka to 19.1 ka (Table 5.7). All sedimentary rocks, including carbonates, rose to 15.4–21% at the same time. First there is a sharp rise of the clastics from 1.5 to 11.8% (of the rock fragments) between 25 and 24.6 ka, which translates into an eight-fold increase of the clastics in response to ice advancing onto the Norwegian shelf. Glacial erosion of the outer Norwegian shelf accounts for one-seventh to one-fifth of the IRD from 24.6 to 19 ka. Therefore, the maximum percentage of shelf-derived sedimentary IRD is 7.4% overall, and 21% of the ice rafted rock fragments (without quartz and feldspar).

Consequently, any sample with clastic sedimentary rocks in excess of 7.4% of the total IRD requires an additional source outside Norway and its shelf. Since central East Greenland and the Wandel Sea Basin in Northeast Greenland are already ruled out for reasons discussed earlier, we are left with Svalbard and the Barents shelf, Franz Josef Land, and North Greenland as the most likely sources. Of those, North Greenland has been detected as a weak influence only for the western Norwegian Sea (Lower Paleozoic acritarchs), while Franz Josef Land seems to have contributed little, based on the absence of basalts with lithologies matching those in source area samples from there.[20] This leaves us with only Svalbard and the Barents shelf as the most likely source for samples with more than 7 or 8% clastics. The amount of clastic sedimentary rock IRD increased abruptly after 18 ka when the Barents Ice Sheet disintegrated (Bischof, 1994; Elverhøi et al., 1988; 1990; 1993; Weinelt et al., 1991). The southward flow of surface waters and icebergs from the Barents shelf produced a coarse layer in sediments of the Norwegian Sea that can be traced from directly off the Barents shelf (Weinelt et al., 1991) to the northwestern Vøring-Plateau (cores 23055, 23056, 23057, 16052; Bischof, 1994) in the Norwegian Sea.[21] On the northwestern Vøring-Plateau, clastic sedimentary rock percentages among > 500 μm IRD rose abruptly from less than 10 to over 80% at approximately 17 ka (Fig. 5.21). The clastics are uniform, lithologically identical, and were partly put into the same clusters as those probed in samples from the Barents shelf (23258, 23259, 23261 and 23262). As an example, look at Table 5.4 (cluster 7), which unites a sample from the Barents slope (23259, 129–130 cm) with five samples from the Norwegian Sea on the basis of their close mutual similarity. One of them is box core 23055, 30–31 cm, with an age of 16700–16300 years bp. The other four belong to oxygen isotope stages 8, 6 and late 1 (very recent). The clastic sedimentary rock IRD peak of GKG 23055 (Fig. 5.21) is no isolated ocurrence, but can be seen in other cores from the Vøring-Plateau as well (Fig. 5.22).

But at KAL 23074, close to the Norwegian continental margin, the southward flow with clastic IRD-bearing icebergs must have interacted with northward flowing currents that carried icebergs loaded with crystalline rocks from coastal Norway. Between those two opposite currents, eddy diffusivity led to mixing, thereby reducing the percentages of clastic rock IRD at this site to 30%. Between roughly 17 and 12 ka, IRD off the Norwegian continental shelf was a mixture of a northern source (Svalbard and the Barents shelf) and material from western and southern Norway. Further to the northwest, the IRD initially, at 17 ka, was almost pure

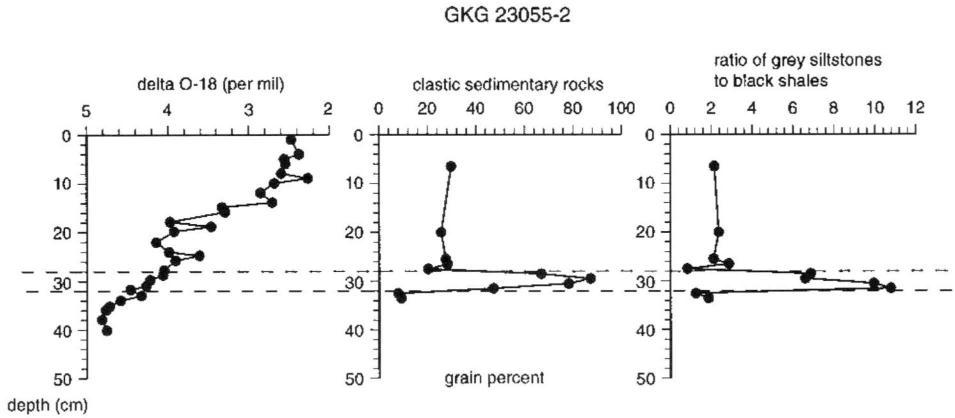

Fig. 5.21(a). Delta O-18 (left), percentage of clastic sedimentary rocks among > 500 μm IRD (middle), and the composition of the clastic rocks, expressed as the ratio between grey siltstones to black shales, for box core 23055-2. Note the drastic rise of clastics at 31 cm, at 17,650 calendar years bp. The clastic rock peak occurred during the initial rise of delta O-18 values between 18 and 14.6 ka. This iceberg pulse came from the Barents Ice Sheet, based on percentages, grain petrographies (Bischof, 1994), cluster and factor analysis (this book), and matching palynomorph assemblages (Bischof et al., 1997). The ratio of grey siltstones to black shales shows how petrographically different the Barents shelf rocks are from those transported by ice before and after the decay of the Barents Ice Sheet. For comparison, clastic rock IRD in the northern Greenland Sea–Fram Strait region has conservative ratios around 1.

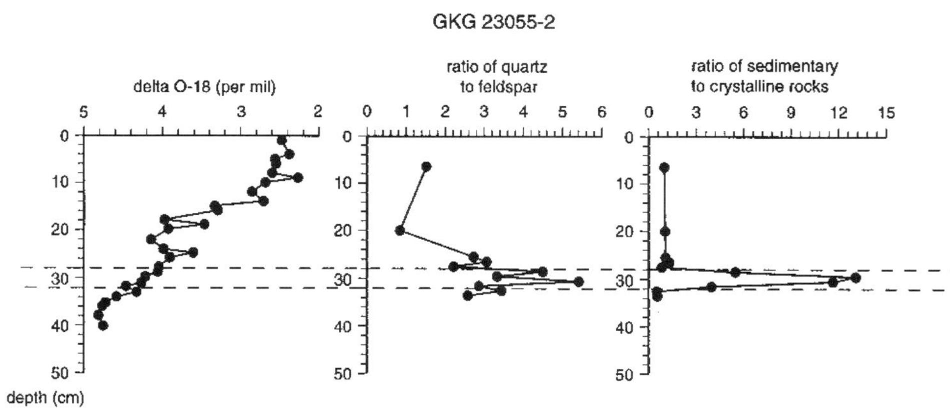

(b) The ratio of quartz to feldspar in GKG 23055 also responds to the change of the IRD composition and rises to values of more than 5. This change underscores further that a very profound change in the direction of IRD delivery, and thus the iceberg drift, had occurred.

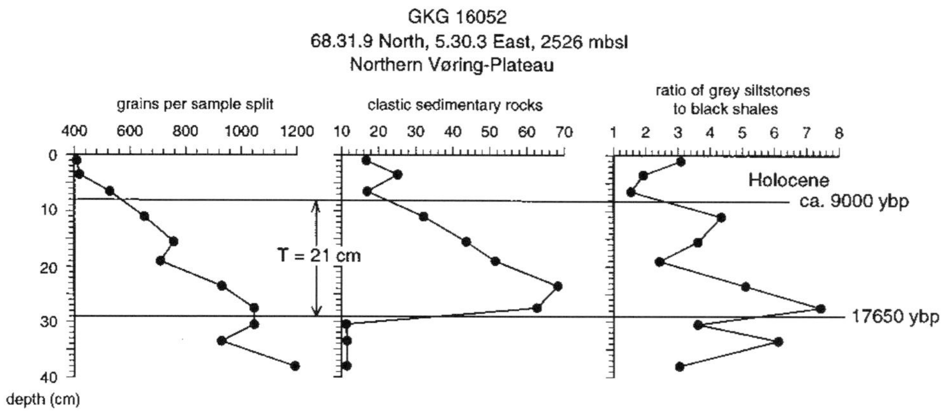

Fig. 5.22. Percentage of clastic sedimentary rocks among >500 μm IRD, the ratio of grey siltstones to black shales, and the number of grains counted in GKG 16052, a core from the northern Vøring-Plateau. The disintegration of the Barents Ice Sheet again shows beautifully as a sharp rise of both the percentage of clastic rocks among the IRD and the ratio of grey to black clastics at the same depth (30 cm) as in GKG 23055. Note that, in contrast to the GKG 23055 record, the clastic rock peak does not reach quite as high a maximum value, and the peak is more drawn out, much broader. Here the thickness is 21 cm, about four times as much as in GKG 23055. Also, the peak lasted from 17,650 cal. years bp until the Pleistocene–Holocene boundary, at 9000 years bp.

Barents shelf-derived. At this time, the beginning of the last major deglaciation, the force of southward currents and icebergs was so strong that it effectively removed almost all Norwegian-derived icebergs from this area. Later, as the iceberg sources in the Barents Sea became exhausted, the northward currents became stronger and initiated the change to present-day conditions (Fig. 5.23).

Fig. 5.23. Ice drift and surface circulation in the Norwegian Greenland Sea at the time the Barents Ice Sheet (BIS) disintegrated, between 18 and circa 12 ka. During this 6000-year-long period, icebergs drifted from the western part of the BIS to the eastern Norwegian Sea (thick arrows on the upper right). At approximately 65 degrees north, these southward currents encountered a northward flow of two kinds: one was a coastal current with icebergs and meltwater from the southern part of the Fennoscandian Ice Sheet (FIS, lower right), and the other one was a northern branch of the North Atlantic Drift (bottom and centre). This current protruded northward through the centre of the Norwegian Greenland Sea and split the current with icebergs from the BIS into two halves: one through the Norwegian Sea, and another one through the Greenland Sea.

see overleaf

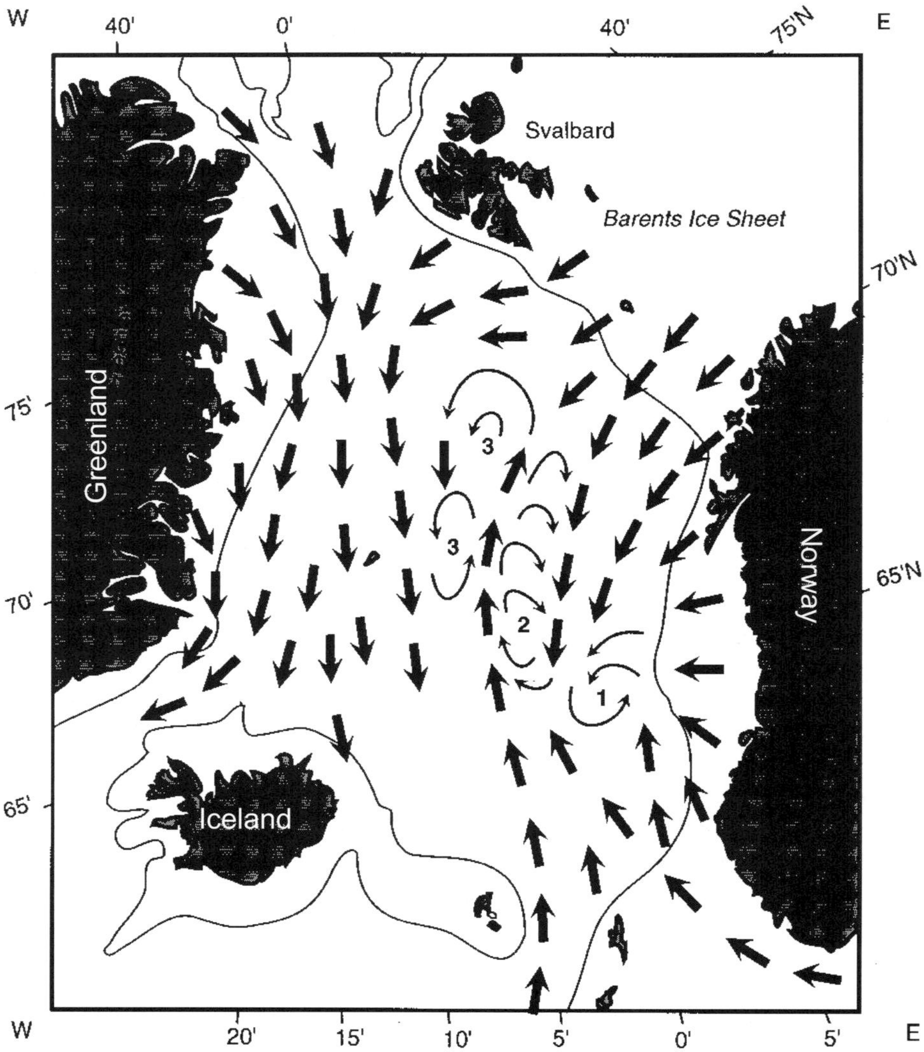

Because of the opposite directions of these currents, gyral, circular motions resulted in between. Three gyres are recogniseable: (1) a counterclockwise gyre in the eastern Norwegian Sea, resulting in a mixture of Barents-derived clastics with IRD of southern provenance containing chalk and flint (as seen in KAL 23074, Fig. 5.19); (2) a clockwise gyre further to the northwest in which mixing between cold, iceberg-bearing waters from the southwestern BIS and possibly warmer North Atlantic waters occurred, and (3) another counterclockwise gyre in the central Norwegian Greenland Sea where waters and icebergs from the northwestern BIS mixed with waters from the North Atlantic Ocean. Icebergs from the central western part of Norway were forced to move southwestwards around the current from the southwestern BIS, then drifted northward into the central Norwegian Greenland Sea where they looped back with gyre 3.

5.6 SURFACE CIRCULATION AND ICE DRIFT DURING THE LATE WEICHSELIAN GLACIATION

From 25 until approximately 19 ka, ice drifted in directions as depicted in Fig. 5.24. This circulation system satisfies all conditions that must be met to produce the observed IRD composition in the Norwegian Greenland Sea. The main elements are:

(1) northward flow in the eastern Norwegian Sea;
(2) southward flow in the Greenland and Iceland seas;
(3) west- and southwestward flow in the northern Norwegian Sea; and
(4) a large, counterclockwise rotating gyre in the Norwegian Sea centred at 65° North.

This flow regime produced ice rafted debris of relatively low petrographic diversity with dominating crystalline rocks from southern Scandinavia and the Baltic region in the southeastern Norwegian Sea, and thoroughly-mixed, very diverse IRD from many different sources along the entire eastern margin of the Norwegian Greenland Sea in the centre of the Norwegian Greenland Sea. The IRD in this central, well-mixed zone contains elements from northern, western and southern Norway along with contributions from as far away as southern Sweden, Denmark, Svalbard, and the bottom of the Barents Sea. In addition, influences from the Tertiary volcanic rocks of the northern British and Faeroer islands are clearly identified, and there may even be a small amount of IRD from the Arctic Ocean. For the westernmost cores in the Norwegian Sea (cores 23059, 23060 and 23062; Fig. 5.1), a partial influx from the clastic sedimentary formations of North Greenland was detected in the form of a peculiar black shale with plant cell structures and Lower Paleozoic acritarchs. This shale is increasingly frequent towards the north and was not found in IRD from east of the 2°W meridian.

The well-mixed character of IRD in samples from the central Norwegian Greenland seas is also the reason why factor analysis produced variable scores for several of the 13 identified source areas (= factors), rather than giving one very high score for a particular factor and low scores for all others.

The major sources of IRD, from the geological interpretation of the factors and the samples' composition, are:

Factor 1: When the factor score is high positive: coarse crystalline granite–gneiss terrain in central western Norway between 63 and 65° North; when the score is very negative: fine-grained clastics from the western Barents shelf.
Factor 2: Mixture of volcanic rocks from Jan Mayen and grey and green quartzites from the eastern Arctic Ocean.
Factor 3: Plutonic rocks from northern Norway and perhaps also from northern Finland.
Factor 4: A gneiss–mafic intrusive rock terrain somewhere in southern or western Norway.
Factor 5: Multicoloured quartzites from the Caledonian nappes of central western Norway.

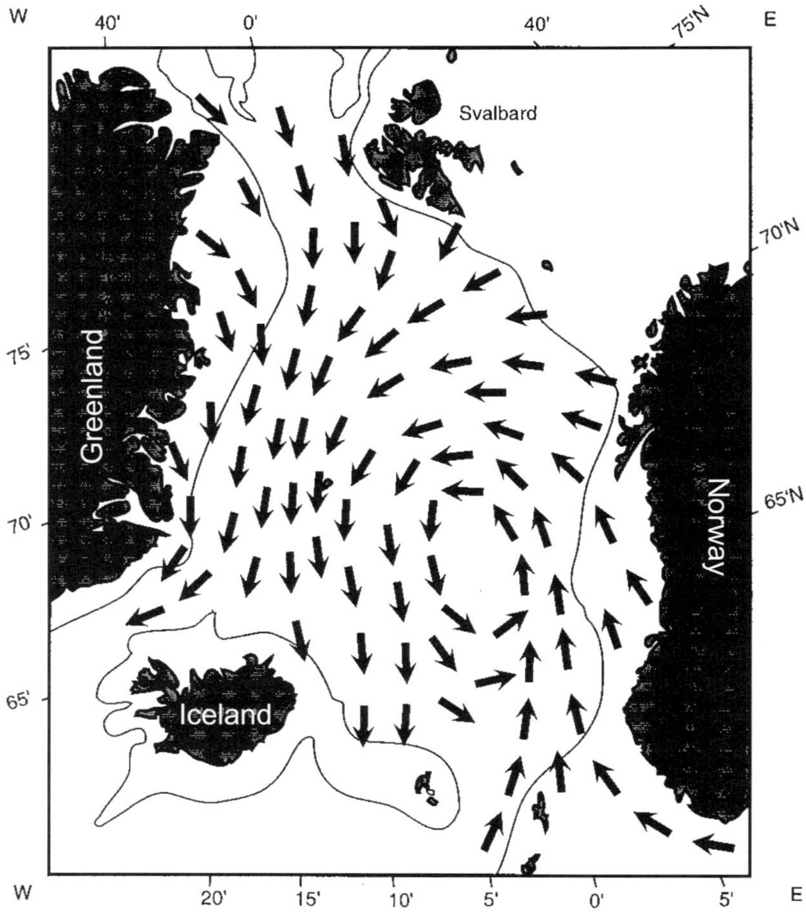

Fig. 5.24. Ice drift and surface circulation during the time period of 25 to approximately 19 ka, the Last Glacial Maximum (LGM). The main elements are northward flow in the east, southward flow in the west (Greenland and Iceland seas), and west- to southwestward flow in the northern Norwegian Sea. Between the north- and southward moving currents, a large counterclockwise gyral circulation in the south-central part of the Norwegian Greenland Sea developed. Within this gyre, thorough mixing of icebergs from southern Scandinavia with icebergs from northern Norway and the western Barents Ice Sheet was possible. The product is highly diverse, well-mixed IRD from many different sources in the central Norwegian Greenland Sea, and lithologically more uniform IRD from fewer sources composed of mostly crystalline rocks in the eastern Norwegian Sea. The circulation shown here was driven by an inflow of waters from the southeast. This inflow had two parts: (1) the eastern side experienced an inflow of waters and icebergs from the northern North Sea and the Baltic Sea, bringing along chalk, flint, and limestone fragments. (2) Between the Orkney and Shetland islands (at the bottom right), and the Faeroer Islands (bottom centre), a northern arm of the glacial North Atlantic Drift moved northwards.

Factor 6: Chalk, flint, limestones and other rocks from southern Scandinavia and the Baltic Sea.
Factor 7: Not interpretable.
Factor 8: Serpentinites, granites and clastics from the inner shelf (?) of western Norway between 65 and 66°30 North.
Factor 9: Not interpretable.
Factor 10: Clastic component of the northern Norwegian shelf north of 66°30.
Factor 11: Phyllites and mica schists from northern Norway, 65 to 68° North.
Factor 12: Black shales and limestones from North Greenland (?).
Factor 13: Coal from the East Siberian Sea.

In summary, the IRD in the Norwegian Greenland Sea is composed of contributions from 13 distinct sources. Of those, seven are different terrains of Norway, and three are sources in the periphery of the Arctic Ocean (East Siberian Sea, North Greenland, and most probably the Taymyr Peninsula). Another important source is the southern Scandinavia–Baltic region. Two of the factors cannot be interpreted as geologic source terrains.

5.6.1 Driving forces

The forces needed to drive the circulation shown in Fig. 5.24 were most likely the same that drive it today, in part oceanic and in part atmospheric. Kinetic energy was provided by an inflow of waters from the northeastern North Atlantic through a channel between the northern British and the Faeroer islands. Another northward flow of water existed south of Norway. These waters came from the northern North Sea and from the deep channels south of Norway and between Norway and Sweden. Both watermasses together formed an ancestor of the Norwegian Current or the Norwegian Coastal Current. The northward movement reached to about 67–68° North where west- to southwestward currents caused a deflection to the left. This created the gyral circulation in the Norwegian Sea. In the Greenland and Iceland seas, conditions were very much as they are today with strong southward flow. The southwestward motion of icebergs (and surface waters?) in the northern Norwegian Sea was most probably driven by transfer of wind energy onto the sea ice surface. The combined Barents and Fennoscandian ice sheets must have had a large pool of very cold, and therefore heavy, air like the Antarctic and Greenland ice sheets have today. The radial flow of air from a high pressure centre over the ice sheet transferred its energy onto the sea surface which was covered by sea ice and icebergs, and the icebergs responded with an offshore drift perpendicular to the margin of the Barents Ice Sheet.

Alternatively, the circulation shown in Fig. 5.24 could have been entirely driven by the atmosphere. All land areas adjacent to the Norwegian Greenland Sea were covered by large ice sheets, and thus two large high-pressure cells existed in the west over the Greenland Ice Sheet, and in the east over the combined Barents and Fennoscandian ice sheets. The resulting air flow from high to low pressure would set in motion a counterclockwise gyre as it is shown in the figure. In this atmospheric-driven scenario, the inflow of North Atlantic and North Sea waters into the

Norwegian Sea would be the passive product of the atmospheric pressure distribution as opposed to actively driving the surface motion.

A third alternative is that the reconstructed ice drift was the product of both atmospheric and oceanic forces. The ocean component would be the inflow of waters from the North Atlantic. Superimposed on this was the atmospheric circulation from two high-pressure areas at the margin towards the centre of the Norwegian Greenland Sea. Fütterer et al. (1988) state that icebergs around the Antarctic continent drift mainly with the Antarctic Coastal Current. Similarly, Johannessen et al. (1994) reported that the velocity and direction of an ice flow equipped with an ARGOS buoy with attached current meters is largely the same as the movement of the waters down to 250 m water depth (Fig. 5 on page 114 of their article). Their observations were made near the ice margin in the Greenland Sea where the ice can float freely. But observational and modelling studies on relatively free-drifting thin sea ice in the circum-Antarctic waters show that 'the mean ice drift is being driven by both low frequency wind and ocean circulation patterns' (Geiger et al., 1998). For these reasons, the ice drift vectors of Fig. 26 most likely represent the long-term vector average of the atmospheric and oceanic circulation.

The delta O-18 values during the period from 23 to 19 ka are uniformly heavy throughout the entire Norwegian Greenland Sea (Sarnthein et al., 1992), which confirms very cold conditions. The Norwegian Greenland Sea was ice covered throughout the year with perhaps some brief opening in the summer months. There is no evidence in the oxygen isotope record of significant meltwater discharges.

5.6.2 Fennoscandian ice sheet dynamics

In addition to allowing us to reconstruct the Late Weichselian ice drift and surface current directions, the IRD composition also contains information about the dynamics of the Fennoscandian Ice Sheet (FIS). As we have seen in Fig. 5.17, six IRD peaks with sharp lower and upper boundaries occurred within a 5000 year long period of prolonged very cold conditions and oxygen isotope values of over 4.5 per mil. The ice rafting peaks therefore are not the product of temperature changes in the seawater, because that should have led to lighter isotope values, and they in turn also have not produced significant quantities of meltwater, which also would have the same effect on the oxygen isotope signature as a temperature change. The IRD peaks are independent from a temperature signal and thus are purely and entirely the result of physical disintegration of specific sub-sections of the FIS. The general theory according to MacAyeal (1993) is that it takes time to build up sufficient amounts of ice before an ice sheet develops a profile that forces the ice to flow and to discharge icebergs to the ocean. Relatively long periods of ice buildup are followed by comparatively short iceberg pulses during which large quantities of icebergs are being disgorged. In the glaciomarine sediment record, this time sequence of events leads to a sediment sequence in which distinct, sharp peaks of coarse IRD are separated by fine-grained sediments whose terrigenous components consist of slowly settling clay and silt from distant sources. Such sequences are dominant in the North Atlantic south of the Iceland Fearoer Ridge and have been described thoroughly by Ruddiman (1977), Heinrich (1988), Bond et al. (1993), and Bond and Lotti (1995).

But the situation was different in the Norwegian Sea, where ice rafting was continuously operative throughout the Late Weichselian. Some coarse-grained ice rafted debris was always dropped from drifting icebergs, either by falling from the bottom of the icebergs, or by debris-laden ice being washed overboard during the brief summer melting season (see colour section). Because of the continuing cold conditions, the IRD peaks of core 23074 directly translate into the number of icebergs that drifted at any given time in the Norwegian Sea.

Of the 11 samples shown in Table 5.7, five represent IRD peaks and thus times of massive iceberg calving under cold conditions. The lowest sample of core 23074 represents a moderate IRD peak and a meltwater or temperature peak. The icebergs at this time (25.6–25 ka) came from the southern part of the FIS and from some of the volcanic islands in the southeastern Norwegian Sea. The next sample upsection, sample 10, represents a high and sharp IRD peak at 24.6–24 ka. Compared to the sample below, the IRD is more diverse, with 35 as opposed to only 27 IRD lithologies, and the proportion of sedimentary rocks is eight times higher than before (in percent of the rock fragments). This increase of the proportion of sedimentary rocks demonstrates that the FIS at this time extended onto the outer shelf where sedimentary rocks occur. The advance occurred between 25 and 24.6 ka. The sample contains chalk and limestones, and thus evidence of a partial derivation from the Baltic region. Meta-arenites and -argillites, which were almost missing in sample 11, are nine times more abundant in sample 10. Thus, the iceberg source area was much broader from 24.6–24 ka than before, extending southeastwards into the Baltic region and northward to central western Norway. On the other hand, the source that previously provided the dark green basalts, most likely one of the Orkney, Shetland or Faeroer islands, was no longer a significant contributor. Sample 10, similarly to sample 11, was an IRD peak accompanied by a meltwater oxygen isotope peak (Fig. 5.17).

Upsection, sample 9 is the first one with an IRD peak that was deposited under maximum cold conditions. It is less diverse than the previous sample, which shows that the icebergs at this time (23.4–23 ka) came from fewer sources. The column with granites, diorites and other intrusives is almost empty (Table 5.7) and shows that the terrain that consisted of various types of intrusive rocks was no longer iceberg productive. The sample contains a relatively complete set of sedimentary rocks, including Baltic lithologies, grey, silver and green meta-argillites, and dark and light gneisses in equal abundance. The latter observation is very rare, as light coloured gneisses are usually much more abundant than the dark types. Therefore, at the period from 23.4–23 ka, the icebergs came from the western and southern part of the FIS, which at this time already extended onto the outer shelf. Sample 8 (21.6–21.3 ka) is even less diverse than sample 9 and contains various sedimentary rocks, but without the red and brown siltstones that were present in sample 9, containing instead brown clay-ironstones. Since red and brown clastics are most abundant in a dredged sample from the northern Norwegian shelf (sample 16327), the presence of red and brown siltstones could indicate a derivation from the Norwegian shelf north of 67°N. Sample 7 further upsection (20.9–20.6 ka) contains the full range of clastics, again with red and brown siltstones, and relatively high amounts of white quartzites which were missing in sample 8.

The subtle changes of the IRD composition from one sample to the next are shifts in the primary iceberg sources through time. At times the icebergs came more from an area where metapelites occur, for example between 20.1 and 19.8 ka (sample 6) when phyllites and mica schists are at their maximum, 7.3% of the rock fragments. At other times the sources were located over bedrock with higher concentrations of quartzites such as between 20.9 and 20.6 ka (sample 7), when quartzites accounted for 8.7% of the rock fragments. The relative proportions of meta-arenites to meta-argillites is in fact a useful measure for a shift of the source areas towards the north (where the metapelites occur) or towards the south (where the quartzites occur in the Caledonian nappes). In sediments older than 16.8 ka, quartzites can be up to three times more abundant than metapelites (sample 10), or metapelites can be up to nine times more frequent than quartzites, as is the case in sample 8. However, this changes dramatically between 16.8 and 13.7 ka, after which quartzites outnumber the metapelites by factors of 5 to 20. Overall, however, gneisses are the most important IRD lithology during the Last Glacial Maximum, and therefore, the FIS on western and southern Norway produced the majority of the icebergs from the Norwegian Sea.

5.7 A PALEOCEANOGRAPHIC MODEL

Here the evidence presented thus far will be combined into a model of the Norwegian Greenland Sea at the time of the LGM, circa 25–18 ka, prior to the Barents Ice Sheet decay. The uniformly cold waters moved in a circular fashion, northward in the Norwegian Sea, and southward in the Greenland Sea. The driving forces were the inflow of North Atlantic waters through the southeastern Norwegian Sea, and the atmospheric circulation from the two major ice sheets on both sides of the Norwegian Greenland Sea. Of these two forces, only the inflow from the North Atlantic could have had a warming effect. The difference between an ice age-style, cold Norwegian Sea and the presently temperate, ice-free edition (in addition to orbitally-induced temperature changes) is the difference in the strength of the forces of warming from the strength of the forces of cooling.

Even during the greatest depths of the last ice age, a form of North Atlantic Drift existed. Waters with an ultimate origin in the Gulf of Mexico moved north- and northeastwards across the North Atlantic. That this flow direction existed is documented in the work of Bond et al. (1993) and other authors who analysed the North Atlantic Heinrich Layers. The difference to today's scenario was that upon its northward passage, the ice age North Atlantic Drift faced a steeper temperature gradient, probably resulting in more rapid cooling of the surface waters than today. In addition to this, the ice age global ocean was more saline than today, because the equivalent of 130 m of freshwater were taken out of the global ocean and stored as ice in the Pleistocene ice sheets. If we calculate with an average depth of the global ocean of 3800 m and a present-day salinity of 36.6 per mil, then the Pleistocene global ocean had an average salinity of 37.9 per mil.[21] During its northward transit and gradual cooling, the North Atlantic waters would at one point become heavier than the surface waters of the Late Weichselian Norwegian Sea, which must have been cold and somewhat less saline because of the icebergs from the FIS. The North Atlantic waters would sink and continue their passage at a sub-surface level, no longer influencing the surface water motion and ice drift. I suggest here that this vertical descent occurred during the LGM at the centre of the gyral circulation shown in Fig. 5.24, at approximately 64 or 65° North. At this location in the Norwegian Sea, the physical forcings of the surface circulation changed from primarily oceanic driven (southern Norwegian Sea) to atmospheric driven (northern Norwegian Sea and Greenland Iceland seas).

But this type of circulation had a twist. Because Spielhagen (1990) found chalk fragments in the Fram Strait during the coldest phase of the last glaciation, at about 25–18 ka, the northward surface circulation in the Norwegian Sea at times must have extended much further north and reached the Fram Strait. In my view, only mixing with freshwater could convert a northward sub-surface current into a surface current. In order to understand how this worked, we must turn our attention to the Greenland and Iceland seas. In this region, the flow of surface waters, sea ice and icebergs was generally southward, into the northwestern North Atlantic. There, because of approaching warmer waters, much of the ice melted, thereby lowering the surface salinity of the North Atlantic waters and amplifying their cooling. Depending upon how much of the ice melted, and to what extent the freshwater was absorbed into the surface mixed layer, the density of the North Atlantic surface waters could decrease to a level where they would no longer be heavier than the glacial surface waters of the Norwegian Sea. Consequently, they would continue to move northward at the surface and not at a deeper level.

Therefore, a pronounced northward extension of the ice drift in the Norwegian Sea was caused by absorption of freshwater from melting icebergs in the northwestern North Atlantic. And the absorption, in turn, depended on the amounts of icebergs drifting southward into the northwestern North Atlantic, and on the rate of diffusive mixing between the fresher and saltier waters. During times when the iceberg drift was low, the North Atlantic inflow into the Norwegian Sea was saltier and heavier, and the waters would descend and allow a surface flow as shown in Fig. 5.24. But when large quantities of icebergs drifted into the northwestern North Atlantic, the surface current became less saline, lighter, and continued to move northward up to Fram Strait.

Figure 5.25 shows what I like to call the 'long loop' in which icebergs from the Baltic region could reach Fram Strait. The loop forces surface waters and ice of the Arctic Ocean, the Norwegian Greenland Sea and the northern North Atlantic to communicate with each other. Further credence for this model is provided by the already reconstructed flow directions in the North Atlantic (Ruddiman, 1977), and by the composition of IRD in the very coarse Heinrich Layers. They contain abundant carbonate IRD from the eastern Laurentide Ice Sheet (eLIS), but towards the northeastern North Atlantic, the IRD composition changes to a mixture with a lot of non-carbonate lithologies whose origin thus far is unknown (Bond et al., 1993). It is very likely that this non-carbonate IRD came from East Greenland and perhaps also partly from the Arctic Ocean. Volcanic rock fragments and volcanic glass in this IRD clearly demonstrate the influence of Iceland.

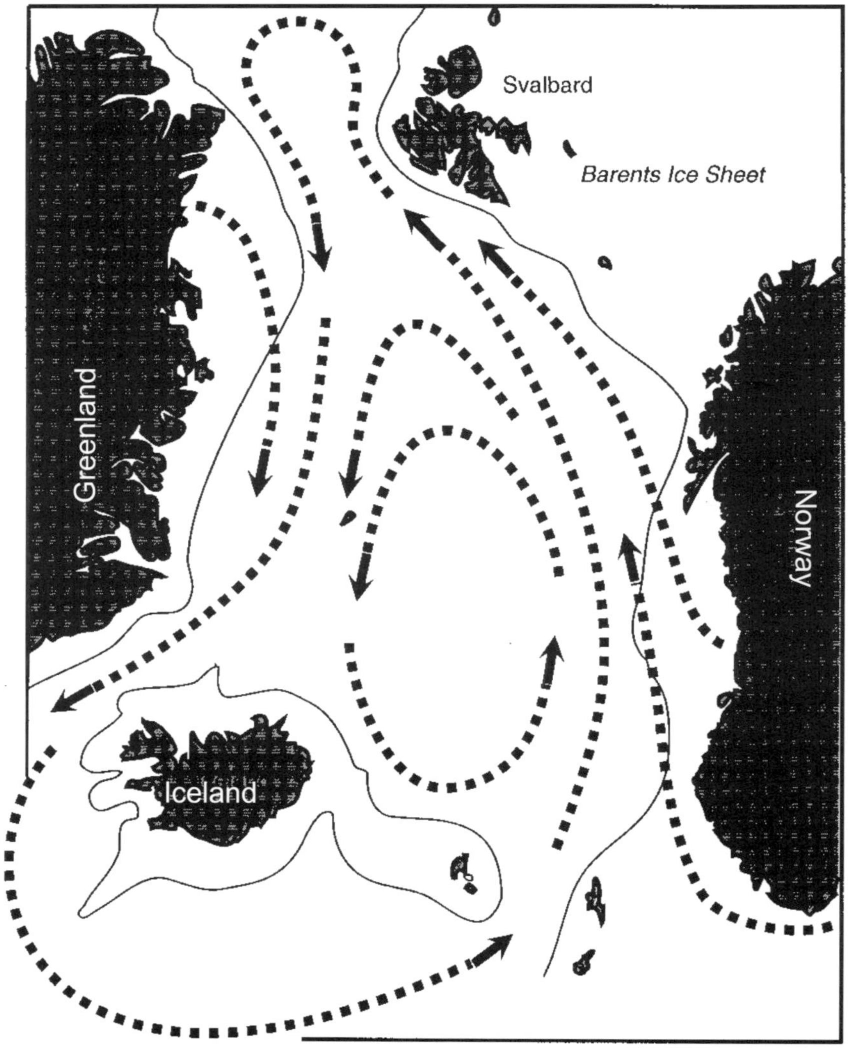

Fig. 5.25. The long loop. In this scenario, massive amounts of icebergs from the Arctic Ocean and Greenland drifted into the northwestern North Atlantic (left side of the diagram). The melting icebergs mixed with the northern branch of the North Atlantic Current, and the resulting lower saline and relatively light surface waters flowed around Iceland and the Faeroer Islands back into the Norwegian Sea (bottom). Icebergs from southern Scandinavia and the Baltic region with chalk joined the flow (lower right) and were able to reach Fram Strait under these conditions. The overall picture is a long, closed loop between the Arctic Ocean, the Norwegian Greenland Sea and the North Atlantic, in which relatively low saline and cold waters with abundant icebergs drifted in a counterclockwise fashion.

The long loop alternated with the short loop of Fig. 5.24. Which of these two scenarios dominated at any given time depended on the amount of freshwater-equivalent in the northwestern North Atlantic, which in turn was governed by the amounts of icebergs from Greenland, Iceland, possibly also the Arctic Ocean, and contributions of meltwater and icebergs from the eLIS.

5.8 THE BARENTS ICE SHEET DECAY

The decay of the Barents Ice Sheet and the change of the surface circulation that it caused is described in detail in Bischof (1994). I will not reiterate the content of that article here. Rather, a brief description of the decay of the BIS will be given, and then a few aspects that have not been published thus far will be discussed.

The most drastic change in the IRD composition seen in any Norwegian Greenland Sea core is the switch from primarily crystalline IRD prior to 18 ka to fine-grained clastic sedimentary rocks thereafter. This change was caused by the decay of the Barents Ice Sheet. The existence of a large, partly marine-based ice sheet in the Barents Sea has long been inferred from seismic features on the sea floor of the Barents Sea indicative of ice movement (Elverhøi et al., 1990, 1993), by postglacial emergence patterns (Forman et al., 1995), and from irregularities in the Earth's rotation (Peltier, 1988). However, clear geologic proof was only recently obtained from cores from the western slope of the Barents shelf (Sarnthein et al., 1992; Weinelt et al., 1991). These cores penetrated a layer of variable thickness with abundant clastic sedimentary rock IRD which matches precisely the bedrock on the bottom of the Barents Sea (Elverhøi et al., 1988). Thanks to high sedimentation rates and the continued presence of datable microfossils, the timing of the Barents Ice Sheet decay could be constrained.

Abruptly, within less than 1 cm core depth, the formerly dominant crystalline IRD lithologies (granites, gneisses, quartzites, phyllites, mica schists, etc.) almost completely disappear, and are replaced by a rather monochromatic assemblage of grey silt- and sandstones (Fig. 5.21). In core 23055 from the northwestern Vøring-Plateau, the entire clastic sedimentary rock-dominated interval is only 5 cm thick. At the neighbouring core 23056, the same interval is at 11–16 cm depth, and at 23057, it is at 11–20 cm depth. The maximum clastic sedimentary rock percentages are found in core 23055 with 87.1%. This core also has the most dramatic shape of the clastic IRD peak with both sharp, abrupt lower and upper boundaries. The same abruptness is also seen in other IRD parameters such as the ratio of grey siltstones to black shales and in the ratio of quartz to feldspar. GKG 23055 contains the purest Barents shelf-derived IRD. Maximum percentages of clastics in IRD directly to the west of where the BIS margin was located at 18–17 ka, are in the order of 60–97%.

The interval of highest clastic rock percentages in GKG 23055 has an age of 17,650–17,210 calendar years bp. It lasted only 440 years. During this brief period, very coarse ice rafted sediment from the Barents shelf was deposited at a sedimentation rate of 9.1 cm/ka, approximately six to seven times more than during the time thereafter (LSR = 1.48 cm/ka). This sedimentation rate is also much higher than the relatively conservative Late Weichselian average of 1.7 cm/ka (Henrich et al., 1989). Undoubtedly, massive amounts of icebergs from the retreating BIS arrived at the Vøring-Plateau at this time. The enormous sedimentation rate increase could have simply been the product of an equivalent increase in the number (or density) of icebergs in this particular part of the Norwegian Sea. In this case, there must have been six to seven times as many icebergs per unit surface area than before. But since it is well known from sediment trap experiments that the rates of sedimentation under ice covered waters are highly sensitive to the surface water temperature (Hebbeln and Wefer, 1991[22]), the sharp sedimentation rate increase after 17,650 ybp could be due to the combined effect of the melting of large volumes of icebergs upon encountering northward-moving warmer waters. The scenario must have been as depicted in Fig. 5.23, with a southward current containing icebergs from the BIS sandwiched between several branches of northward moving currents. These northward currents were warmer and caused intensive melting. This is supported by oxygen isotope values of foraminifera that live in the surface waters. For the time slice 18 to 17 ka, their values are relatively heavy near the BIS margin, but lighter at some distance to the west (Sarnthein et al., 1992). This indicates that melting occurred after the icebergs drifted into somewhat warmer waters. The area where these lighter values were found coincides with the area marked '3' in Fig. 5.23.

The trail of icebergs from the BIS leads southward from the Barents shelf margin to the Vøring-Plateau. Its core area is at the location of GKG 23055, at approximately 68° North and 2° West. From this core zone with relatively undiluted Barents-derived IRD, the clastic sedimentary rock proportions decrease towards the east, west and south as a result of mixing with icebergs from other, non-clastic, sources. In core 16052, the clastic IRD peak also has a very sharp lower base, but the values gradually decrease after a peak of 70% (Fig. 5.22). The peak in 16052 differs in two aspects from the one at 23055 (and also the adjacent cores 23056 and 23057): the maximum clastic rock IRD percentages are lower (only 70% as opposed to 87%), and the peak is much broader. The thickness in 16052 is 21 cm, approximately four times thicker than at 23055. The same pattern is also seen at the central Vøring-Plateau core 16069 (Fig. 5.26). Both clastic sedimentary rock peaks resemble each other like siamese twins: they share the same steep beginning followed by a more gradual decline, the same thickness, and the same ratio of grey siltstones to black shales.

The regional decline of the maximum clastic sedimentary rock percentages is strongest towards the south. Of the original 87% at the site of GKG 23055, only 55% are left 100 km to the south at 16069. Further south, to the sites of KALs 23071 and 23074, the percentages drop to values of 30–35%. Towards the Norwegian continental margin (GKG 16061) the values drop to 14.6% on average, at a variability of 11.3–17.4% (Fig. 5.27). This core shows how the percentages of certain IRD lithologies changed on at least three different time-scales: chalk and flint as indicators of a southern IRD derivation show two pronounced peaks spaced more than 10 cm apart (Fig. 5.27b), but the non-chalk carbonates which also are derived from somewhere in southern Scandinanvia are in a longer-term decline that extends over 35 cm. In contrast, mafic minerals (Fig. 5.27e) show comparatively abrupt,

The following colour plates show icebergs, in part with heavy debris load. The examples are from East Greenland (Scoresby Sund) and Axel Heiberg Island in the northern Queen Elizabeth Islands. Pictures courtesy of Robert Gilbert.

Also shown is sea ice with sediment accumulations. Pictures courtesy of Julian Dowdeswell.

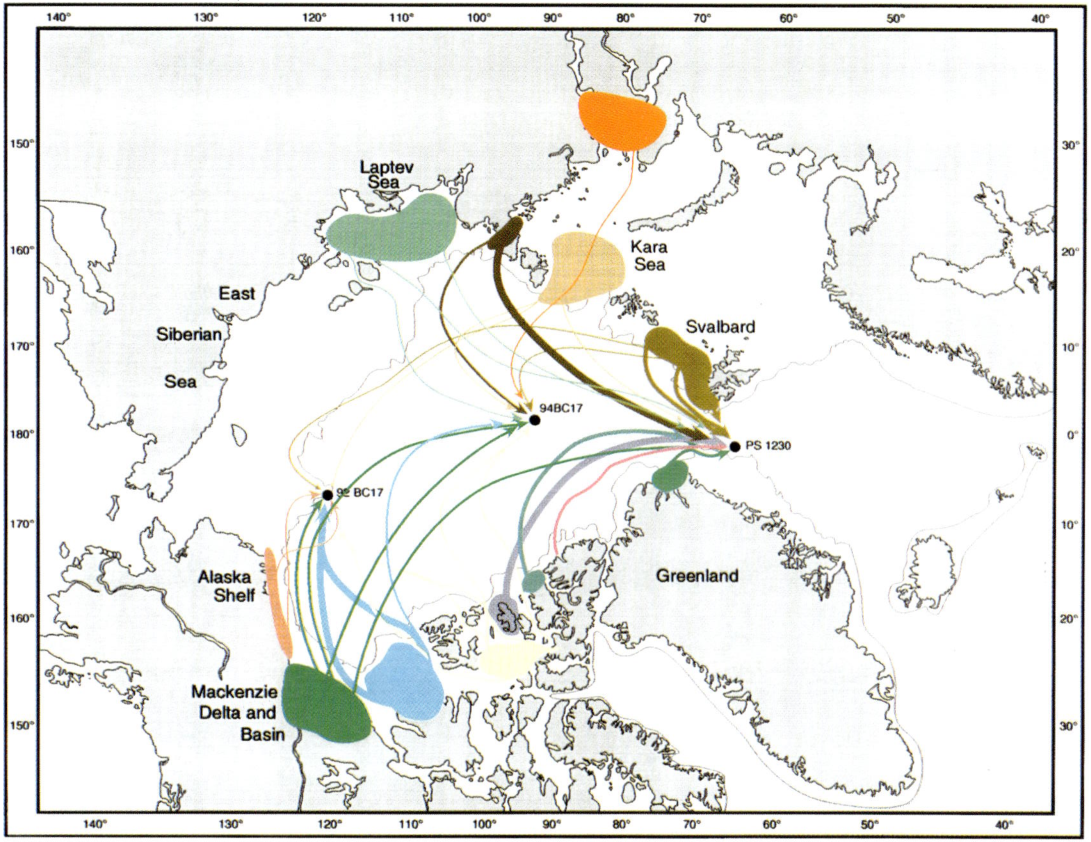

Fig. 6.8(a). Ice drift trajectories as determined by DFA for two cores from the western (92BC17) and central Arctic Ocean (94BC17) and Fram Strait (PS 1230). The time represented by the box core records (approximately) are 0–23 ka (92BC17), 0–45 ka (94BC17), and 0–30 ka (PS1230). The thickness of the trajectories is proportional to the number of determinations from the indicated source. Major sources are the southwestern Canadian Arctic islands, northwestern continental Canada, Svalbard and the Kara Sea. The westernmost core was almost exclusively influenced by the Laurentide Ice Sheet, the central Arctic Ocean received IRD from both sides of the Arctic Ocean, and the Fram Strait core received ice from both the Barents and Kara ice sheets and the Innuitian Ice Sheet. Notably the vast East Siberian Shelf, which was dry land during most of the Wisconsinan Glaciation, delivered nothing.

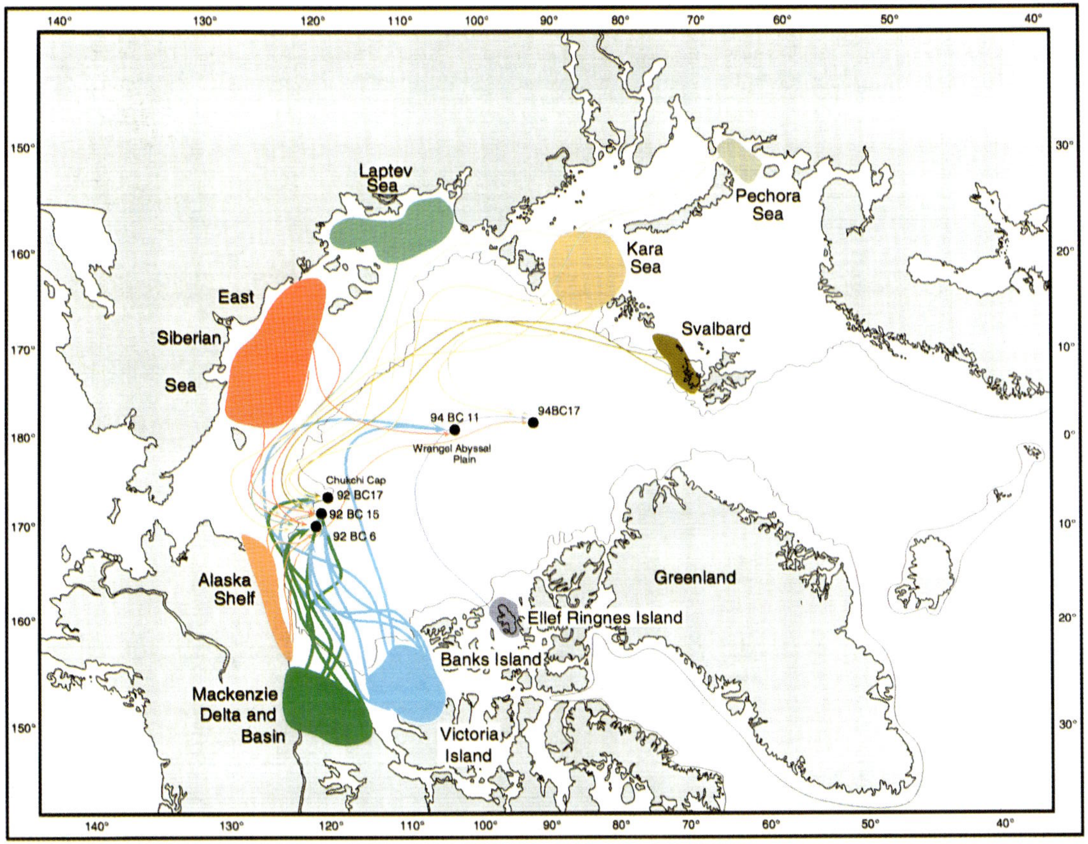

(b)This shows the state of our research one year earlier than Fig. 6.8a. Three cores from the western Arctic Ocean (92BC6, -15 and -17) were heavily influenced by Laurentide sources, except for the Holocene when sources in Siberia, the Kara and Pechora seas became important. This source area change at the end of the last ice age is a clear reflection of the change from glacial iceberg rafting to post-glacial sea ice rafting.

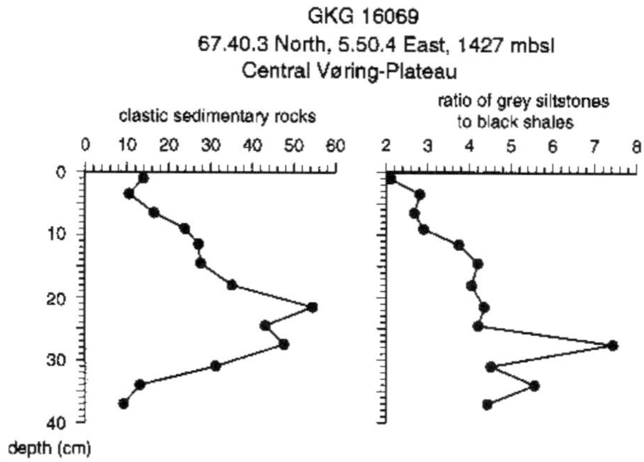

Fig. 5.26. Concentration of clastic sedimentary rocks among IRD >500 μm, and the composition of the clastic rocks in core 16069 from the central Vøring-Plateau. The clastic sedimentary rock peak is very similar to that of GKG 16052. It has a steep lower base and a less inclined upper slope. The maximum clastic sedimentary rock percentage at this site is 55%, still a very high value, but significantly lower than the 87% of GKG 23055, which is located approximately 100 km further north. Therefore the reduction of the clastic sedimentary rock content in the IRD was caused by advection and mixing of icebergs from southern sources into those that came from the BIS.

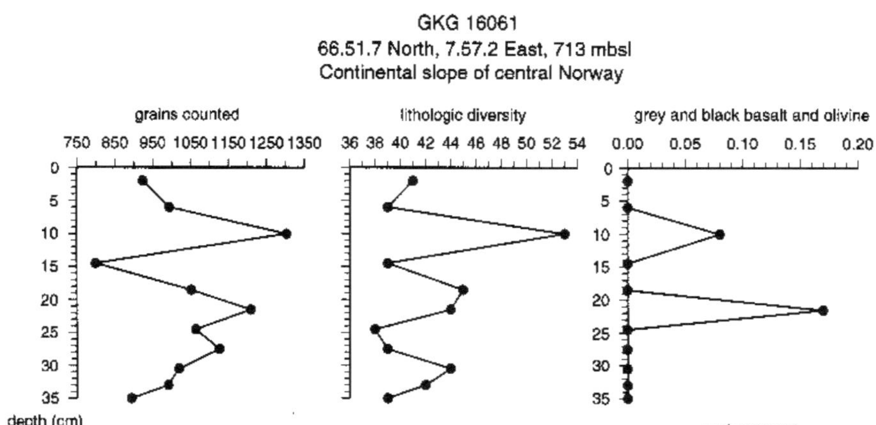

Fig. 5.27. Composition of IRD in core 16061. This core has an upward coarsening sequence as a result of an increasing effect of bottom currents. The top of this core has a Late Weichselian age of approximately 9000 years bp. There is no Holocene sediment. The core demonstrates that the BIS decay had only limited impact at this site, which is dominated by crystalline rocks.

(a) Grains counted, lithologic diversity, and the proportion of grey and black basalt and olivine which characterises debris from Jan Mayen Island. The evidence for eastward currents in this core is very weak.

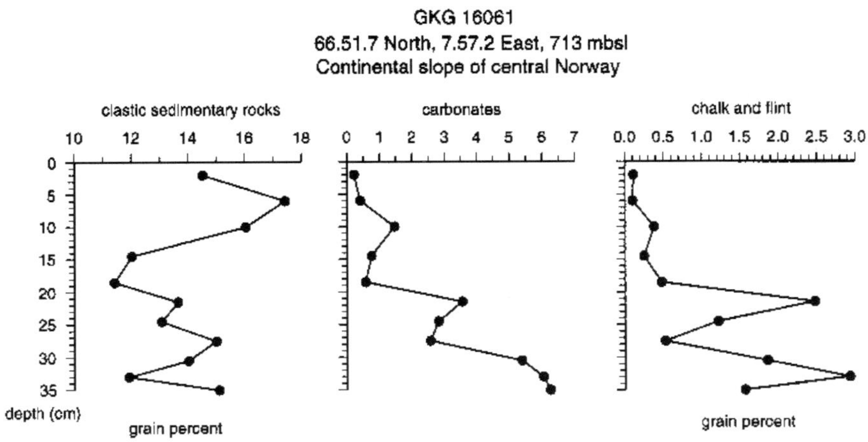

(b) Note the long-term decline from initially over 6% carbonates (from southern Scandinavia) at the bottom to very low values near the top. Chalk and flint have two maxima during which a strong northward flow must have existed.

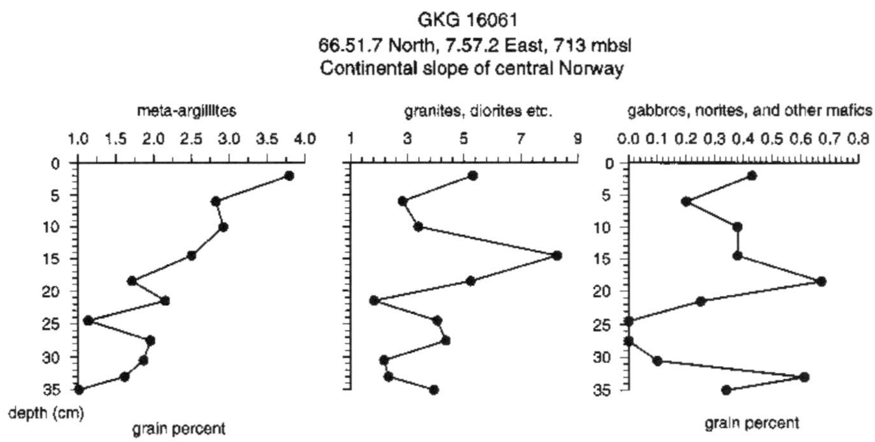

(c) The decline of the carbonates is balanced by an increase of metapelites. Since we have already identified northwestern Norway to the north of Trondheim Fjord as the most likely source of the metapelites, the compensation of dropping carbonate percentages by rising percentages of metapelites shows a gradual northward migration of the iceberg sources. Light and dark coloured intrusive rocks show an independent occurrence pattern and thus are derived from a different source.

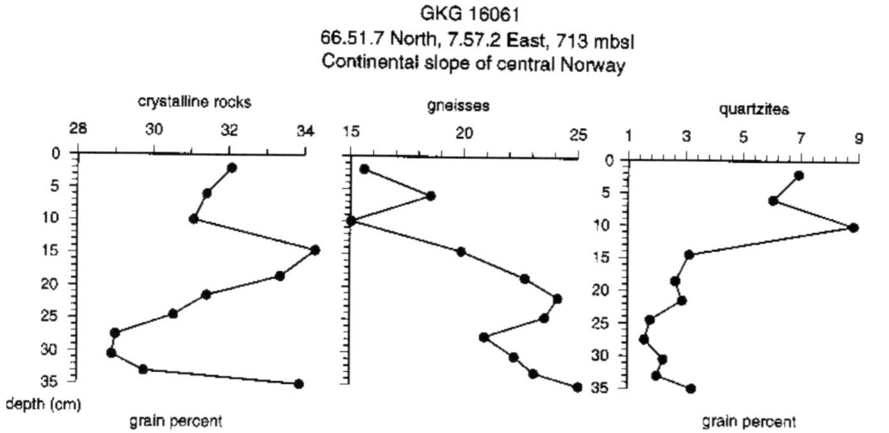

(d) The crystalline rock IRD is not a homogenous group of rocks derived from the same area, but rather consists of diverse rocks from different sources. Note the inverse increases and decreases between gneisses and quartzites upward of 18 cm core depth.

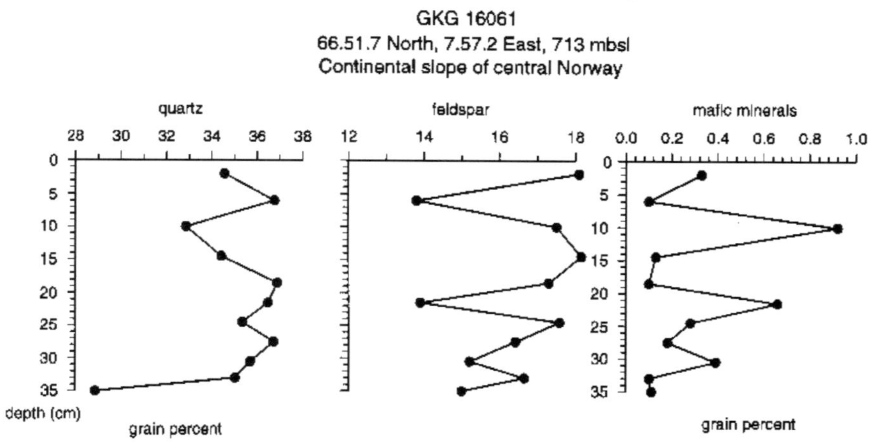

(e) Not only different types of rocks, but also the mineral fragments have experienced significant percentage changes. At this site, quartz is more abundant than feldspar, and mafic minerals (mostly amphiboles and black opaques) oscillate around low values.

sharper changes on depth scales of 2 to 3 cms. Therefore, a long-term shift from a primarily southern to an increasingly more northern (metapelites) source has been overprinted by iceberg discharges from different parts of western Scandinavia. But the influence from the BIS to this particular site was almost negligible. Similarly, another box core from the southern Vøring-Plateau (GKG 16043, Fig. 5.28) also shows only weak influences from the BIS. Its clastic sedimentary rock IRD oscillates between 8 and 20%, with the highest percentages during the Late Holocene, clearly after the demise of the BIS. The dominant IRD lithologies at this site are crystalline rocks from Norwegian sources, plus some chalk and flint from southern Scandinavia.

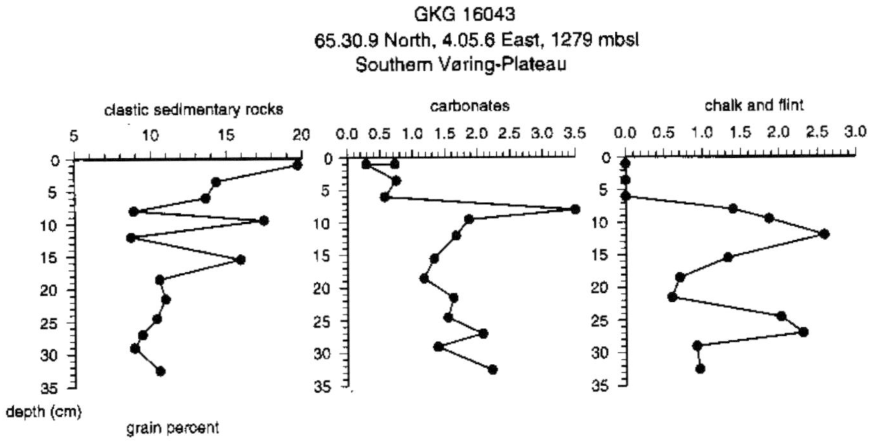

Fig. 5.28. Composition of IRD in box core 16043, southern Vøring-Plateau. (a) Clastic sedimentary rocks are generally in the same range as those in core 16061, indicative of a low influence from the BIS. The highest clastic sedimentary rock percentages are obtained after the end of the Late Weichselian and could not have been the product of the BIS decay; rather they are the product of modern sea ice rafting from Svalbard. Chalk and flint show the same double-peak as in 16061 and denote two periods during which northward currents and iceberg drift dominated this area. After the Holocene began, no more chalk and flint fragments found their way into this part of the Norwegian Sea. Carbonates show one single peak at the beginning of the Holocene.

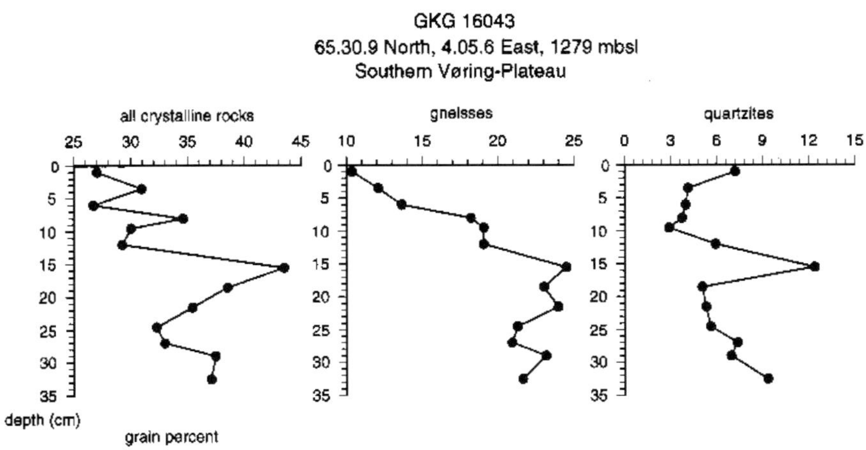

(b) Crystalline rocks, gneisses and quartzites. Note the demise of gneisses at the end of the Pleistocene in the same horizon in which quartzites stay the same.

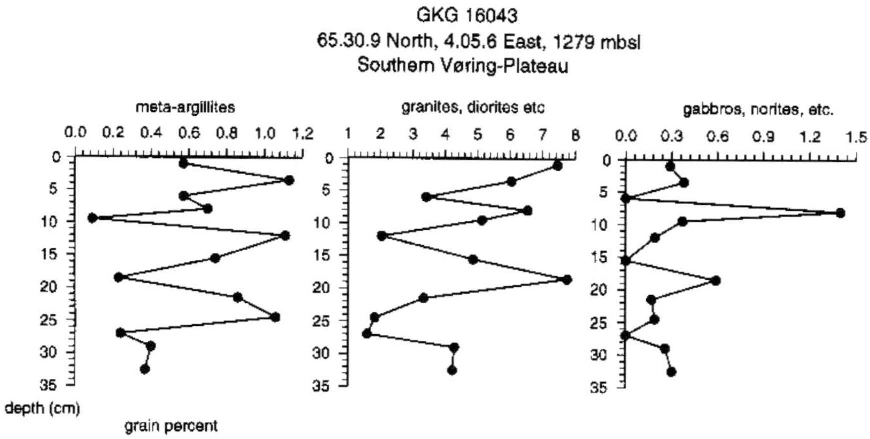

(c) Meta-argillites, light and dark coloured intrusives. Note the inverse abundance patterns between the meta-argillites and granites, diorites, etc. in the Pleistocene section between 25 and 9 cm.

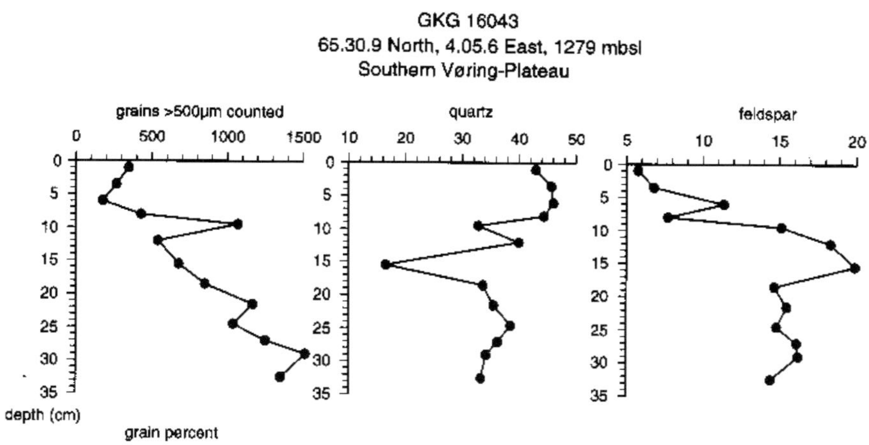

(d) Number of grains counted in each sample and quartz and feldspar content. Quartz averages 34.5% while the feldspar content is lower. In the Holocene section, quartz is five to 10 times more frequent than feldspar, indicative of a shift of the dominant ice source away from feldspar-rich granites and gneisses towards sedimentary rocks. Only once, at 15 cm depth, is feldspar more abundant than quartz. This is the only time that the IRD at this location shares the characteristics of the site of core 23075 and several cores in the central Norwegian Greenland Sea.

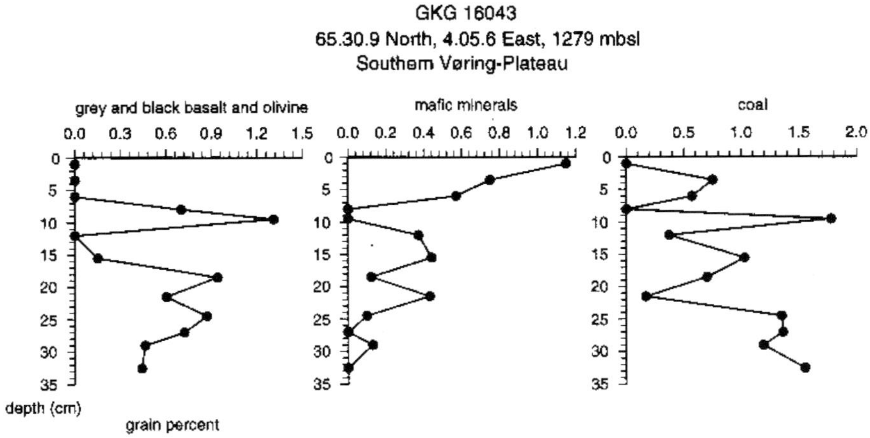

(e) Grey and black basalt and olivine are present throughout most of the Late Weichselian and indicate that some eastward flow from Jan Mayen Island to the southern Vøring-Plateau occurred. The continued presence of coal demonstrates a northern, Svalbard or eastern Arctic Ocean-derived component. Mafic minerals increase towards the present day.

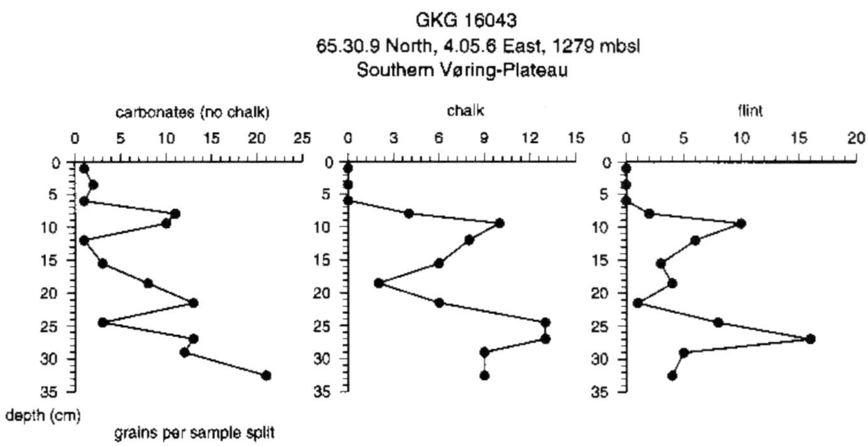

(f) Carbonates (without chalk), chalk and flint as grains per sample split. The parallel shape of the chalk and flint counts demonstrates their common orgin. The long-term decline of the carbonates as seen in 16061 also is seen in this core.

The regional drop of the clastic sedimentary rock IRD clearly shows that the icebergs from the BIS encountered icebergs from other sources with less or even without clastic sedimentary rocks. Since most of these samples also contain indicators of southern provenance (chalk and flint), the icebergs that caused the decrease in the clastic sedimentary rock percentages came from southern and western Norway. Lateral diffusive mixing between the approaching north- and southward currents occurred in the circular mixing zones (gyres) labelled 1 to 3 in Fig. 5.23.

Towards the west of the centre of the southward current (core 23055), the time period of the BIS decay is represented by several samples that integrate across the entire interval, over several thousand years. None of the cores in the west (GKGs 23059–23065; Fig. 5.1) have clastic sedimentary rock percentages comparable to those on the Vøring-Plateau. These cores rather have a complex mixture of mostly crystalline IRD from sources to the southeast of the Norwegian Sea, but they also contain debris from central western and northern Norway (factors 1 and 3, Table 5.5). This mixture was the product of the zone marked with a '3' in Fig. 5.23. These central Norwegian Sea cores contain more feldspar than quartz, which for reasons explained earlier was the result of icebergs from central western Norway north of 65 degrees. These icebergs must have drifted around the south of the current from the BIS. There, the icebergs were caught by the northward moving current in the central Norwegian Greenland Sea.

While the regional percentage changes of clastic sedimentary rocks among IRD are explained by the surface currents, the temporal evolution of the IRD composition (first the sudden jump from low to extremely high clastic rock IRD percentages, and then the following decline) must have a different explanation. The bottom topography of the Barents Sea is characterised by shallow areas and deeper trenches with present-day water depths of over 500 m. One trench is located north of Norway and has a coast-parallel shape, and another one is located south of Spitsbergen. Before the disintegration of the BIS, the entire area was covered by a large ice sheet, whose centre was located east of Svalbard. As the BIS began to recede, the icebergs came from two major calving bays: one was located north of Norway where the bedrock consists exclusively of clastic sedimentary rocks, and another one was located further north, closer to Svalbard. The part north of Norway produced the highest clastic sedimentary rock percentages in Norwegian Sea IRD. As a consequence of this period of massive iceberg rafting, a large part of the ice sheet, the one north of Norway, vanished. This iceberg calving episode split the once joint Fennoscandian–Barents–Svalbard ice complex into two separate units, the Fennoscandian Ice Sheet and the Barents Ice Sheet *sensu stricto*. After approximately 17 ka, the northern FIS could no longer produce abundant clastic sedimentary rocks, because its geometry had changed and the ice from now on overrode both crystalline and sedimentary rocks. The gradual drop of the clastic sedimentary rock percentages in the cores 16052 and 16069 (Figs. 5.22 and 5.26) document this ice margin retreat beautifully.

The remaining ice on Svalbard and the Barents shelf also was in retreat and produced icebergs with clastic sedimentary rocks that contained significantly more black shales than grey siltstones. This type of IRD is seen in the northernmost core from the Barents shelf slope, KAL 23258. Sample 23258, 588.5–591 cm core depth, contains by far the highest percentage of black shales (30.6%), the highest relative proportion of black shales among rock fragments (47.2%), and the lowest ratio of grey siltstones to black shales (0.16) of all Norwegian Greenland Sea cores. These characteristics define the clastic rocks on the bottom of the northwestern Barents shelf. By contrast, the three cores from further south, 23259, 23261 and 23262, contain significantly less black shales (1.2–15.9%), and the ratios of grey siltstones to black shales are among the highest in the entire data set. The samples contain at least three times, and as much as 28 times more grey siltstones than black shales. Therefore, very high quantities of black shales characterise the northwestern Barents shelf, and high proportions of grey siltstones its southwestern part.

The samples with the next highest black shale IRD percentages after 23258 (588.5–591 cm) are Holocene samples from the Greenland Sea, and samples of early Weichselian, stage 6, and stage 8 ages from the central Norwegian Greenland Sea. All these samples have been undeniably affected by icebergs and sea ice from the southern part of the Svalbard archipelago and the adjacent part of the Barents shelf.

With respect of the timing of the ice margin retreat an interesting fact is that another sample of core 23258 from 70 cm deeper in the core contained less black shales (18.7%), and the ratio of grey siltstones to black shales was higher (3.2). At this earlier period, the ice sheet covered a much larger area and eroded not only black shales but various other sedimentary rocks as well. As the ice margin retreat continued, the glaciers vacated the area occupied by the grey siltstones and retreated onto the part of the northwestern Barents shelf that consists of the black shales.

Icebergs from the northwestern Barents Ice Sheet took a westward to southwestward drift track whereas icebergs from the southwestern Barents, respectively the northern Fennoscandian Ice Sheet drifted on a course closer to the Norwegian continental margin (Fig. 5.23). These two essentially southward moving currents were split by a northern branch of the North Atlantic Current. This Weichselian edition became increasingly stronger and broader and later evolved into the modern Norwegian Current–Norwegian Coastal Current system, which keeps the eastern half of the Norwegian Greenland Sea (mostly, as we have learned) ice free.

5.8.1 The Weichselian Glaciation prior to the LGM

Core 23071 extends farther back in time than 23074 owing to lower sedimentation rates. The core has been analysed for chronology with numerous ^{14}C AMS datings which are published in Vogelsang (1990) and Voelker (1999). The bottom of KAL 23071 (730 cm core depth) has an age of approximately 135,000 years. Therefore the core represents the entire Weichselian Glaciation, the preceding Eem warm period (also sometimes referred to as isotope stage 5e), and the last phase of the Saalian Glaciation (European terminology for isotope stage 6; in the US: Illinoian Glaciation). The core has received coarse IRD at semi-rhythmical intervals. Periods with intensive iceberg rafting from the FIS and other ice sheets alternated with more quiet periods when iceberg rafting was minimal (Fig. 5.29). It appears that individual iceberg pulses were short, lasting perhaps only a few 100 years.

The IRD composition has changed considerably during the last 130 ka. Figure 5.30 shows four of the most important IRD components, clastics, carbonates, light and dark coloured intrusive rocks. All components vary in abundance throughout the core, but particularly so during the last 30 ka. One should not forget that the samples represent 10–20 cm thick core segments spaced up to 50 cm apart. Therefore the samples integrate vertically over the sample thickness, and the composition of IRD in between the samples remains unknown. Synchronous ups and downs of

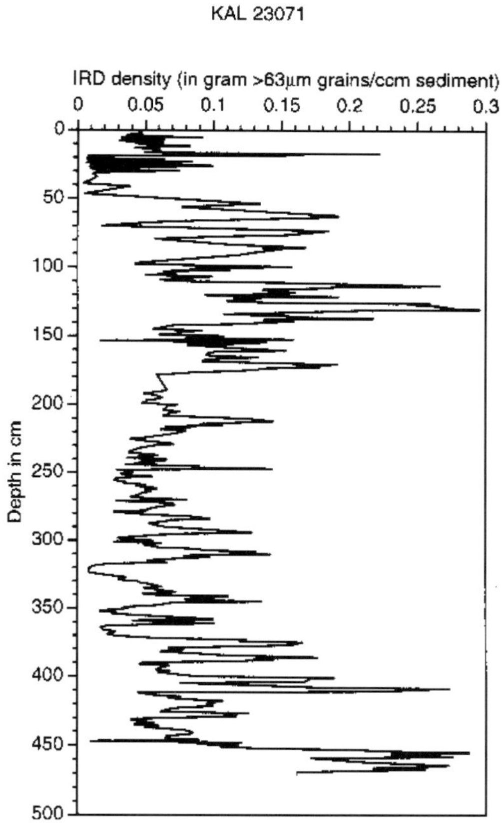

Fig. 5.29. IRD density in KAL 23071, Norwegian Sea. The data are reported as the weight of IRD grains > 63 microns in gram per one cubic centimetre of sediment. The data for this plot were obtained by Elke Vogelsang and kindly provided by Antje Voelker (both at Universität Kiel, Germany). The 470 cm level where the IRD density data end (the actual core is longer) has an age of 60,650 calendar years bp, and the top is recent. The important thing to remember are the numerous IRD peaks of 5–10 cm thickness that follow each other at almost regular intervals. These peaks are periods of intensified ice rafting from primarily the Fennoscandian Ice Sheet.

carbonates and clastics during the earlier part of the core indicate derivation from the same sources. The clastic sedimentary rock percentages during the two peaks at 85 and 135 ka are almost as high as during the last peak which was due to the BIS decay. The BIS thus had a long history: there was an earlier edition during the late stage 6, which in the European terrestrial nomenclature is the Saalian Glaciation (see the chapter on Pre-Weichselian Glaciations in the Norwegian Greenland Sea), and another one during the early Weichselian Glaciation. After 80 ka, the influence from these northern sources gradually diminished due to increasing amounts of crystalline IRD from the Fennoscandian Ice Sheet.

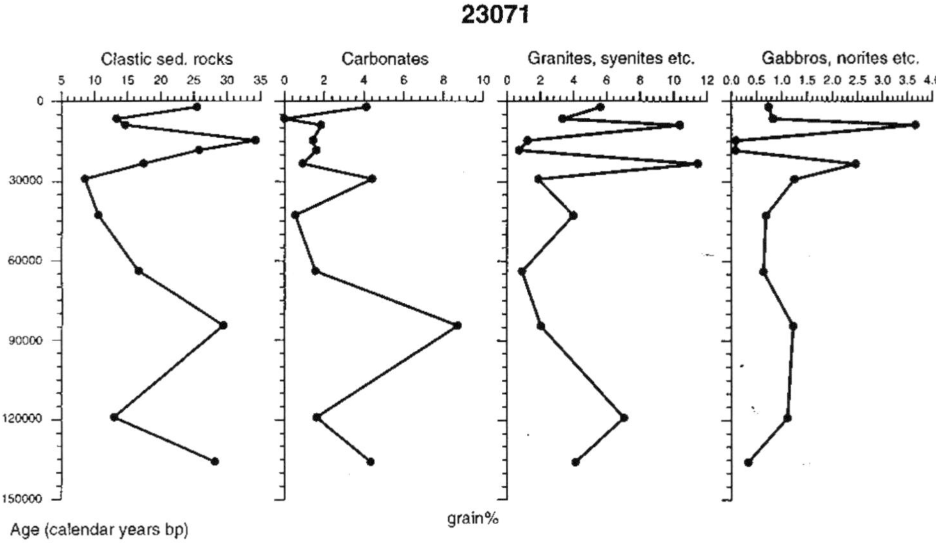

Fig. 5.30. Composition of IRD in core 23071, Norwegian Sea. Shown are four of the most important IRD lithologies. Note the synchronous ups and downs of carbonates and clastic sedimentary rocks between 135 and 45 ka. Clastic rocks are high at the end of the Weichselian Glaciation because of the disintegration of the BIS. Two previous peaks at 85 ka and at the end of isotope stage 6 of almost the same height as the one at 15 ka indicate (at least) two previous editions of the Barents Ice Sheet, one during the late Saalian ice age, and another one during the early Weichselian.

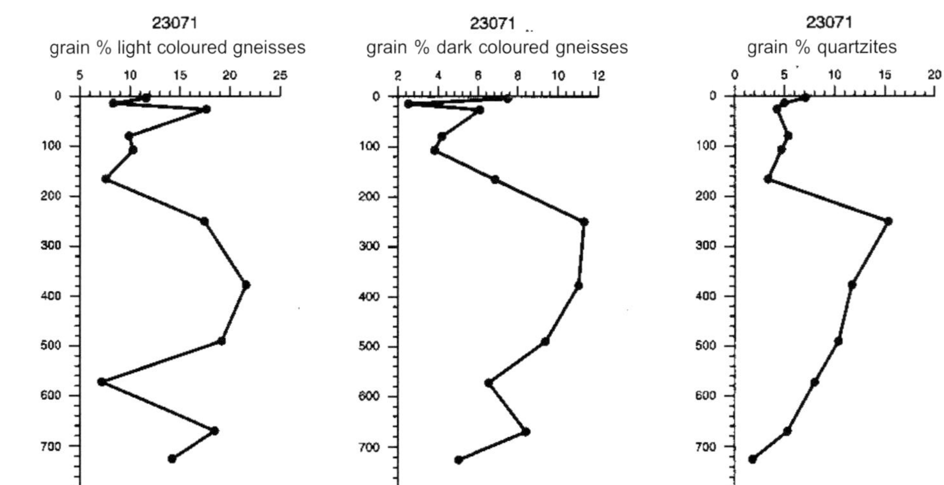

(a) Light coloured gneisses among IRD versus depth in KAL 23071. During the time interval when clastics are relatively low, represented by the 4th, 5th and 6th sample from

the bottom, gneisses are most abundant. This is the shift from formerly clastic rock IRD from sources around Svalbard and the bottom of the Barents Sea to crystalline IRD from Norway and adjacent parts of Scandinavia.
(b) Mafic gneisses essentially replicate this pattern and also are most abundant when clastic rocks are low.
(c) Quartzites are different in that they rise for a long period of time, after which they drop sharply. Clearly, the icebergs came more frequently from an area with metasedimentary rocks, perhaps the Caledonian formations of Norway or the Hekla Hoek rocks of Spitsbergen.

But the crystalline IRD was not a homogenous group of rocks. Its composition overall as well as the proportions of some of their constituents changed. Figure 5.30a, b show that gneisses, regardless whether of light or dark colour, increase and decrease almost synchronously. But quartzites (Fig. 5.30c) show an almost linear, long-term increase that appears to be independent from the gneisses. This demonstrates that not only the proportion of crystalline rocks among IRD changed, but also the internal composition of the crystalline rock IRD. There was a long-term trend from a source terrain composed primarily of gneisses (Archean basement rocks) to another terrain where quartzites are more abundant (Caledonian nappes).

During the early Weichselian, icebergs came primarily from the north and transported a blend of sedimentary and crystalline rocks, matching the bedrock composition of northern Scandinavia and the bottom of the Barents Sea directly north of it, and the Svalbard archipelago. Later, the iceberg sources switched to the crystalline formations of Norway, and within this rather large area they gradually switched from a gneiss terrain to a terrain with quartzites. This succession of different types of IRD suggests earlier ice buildup in the far northern regions around the NGS, which was followed by an expansion of the continental glaciers towards the south. Glacier growth most probably began in Spitsbergen and northern Norway. As these ice complexes, fed by moisture from the NGS, grew in aerial extent and thickness, they coalesced to form the marine-based Barents Ice Sheet in the north, and the Fennoscandian Ice Sheet over Scandinavia with a centre of greatest thickness over northern Sweden.

5.9 SUMMARY OF ICE RAFTING IN THE NORWEGIAN GREENLAND SEA AND AN OCEAN–CLIMATE MODEL

The regional distribution of specific IRD lithologies and their temporal abundance changes allow us to draw the following conclusions. During the Last Glacial Maximum, between 24 and 18 calendar years bp, the general surface ocean circulation and ice drift was strictly southward in the Greenland and Iceland seas, northward in the southern Norwegian Sea, and west- to southwestward in the northern Norwegian Sea. The opposing north- and southward currents interacted to form a large counterclockwise gyre in the Norwegian Sea. This gyre was no steady feature, but changed location and size. At times it formed an elongated north–south trending loop, in whose eastern part icebergs from the Baltic Sea and southern Sweden could drift as far north as to the Fram Strait. This situation is not unlike the present-day arrangement of surface currents, except that during the depth of the last ice age the current was cold and ice covered, not warmer, ice-free water as today. But this northward extension of the gyre was a rare occasion, because its definition is truly based on only two fragments of chalk. Because of the counterclockwise gyre, IRD lithologies, and thus icebergs, from western and southern Norway, Svalbard and the Barents shelf, North Greenland, and places in the eastern Arctic Ocean were mixed together, which led to strongly mixed, heterogenous IRD assemblages from sources several thousands of kilometres apart. In the eastern Norwegian Sea, IRD mostly came from western and southern Norway (south of 66° N) and southwestern Scandinanvia, whereas the central Norwegian Greenland Sea contains increasing amounts of IRD from northern Norway, Svalbard and the Barents shelf. West of a line from Iceland to Spitsbergen, geographically about half of the Norwegian Greenland and Iceland seas, an Arctic Ocean, North Greenland and Spitsbergen-derived IRD signature becomes dominant.

Temporal changes of the IRD composition in the eastern Norwegian Sea were due to sequential discharges of icebergs from different parts of the Fennoscandian Ice Sheet. But the lengthening and shortening of the northward current component in the Norwegian Sea, the difference between the 'short loop' and the 'long loop', was controlled by the density of the inflowing ice-age version of the North Atlantic Current–Norwegian Current system. When waters in the northwestern North Atlantic received and absorbed sufficient amounts of freshwater in the form of meltwater and icebergs from the northeastern Laurentide Ice Sheet, from Greenland, and from the Arctic Ocean, the salinity of the surface waters dropped and the current stayed at the surface, continuing its northward motion with sea ice and icebergs. But when the supply of freshwater subsided, the salinity increased and the current became heavier than the Norwegian Sea surface waters that it encountered. It therefore sank and continued to flow at a sub-surface level, no longer affecting the ice drift at the surface. This situation must have been an analogue to what happens nowadays near Svalbard: there the northward-flowing West Spitsbergen Current cools to a temperature at which its saline waters become heavier than the Arctic surface waters. Therefore, the current sinks and slides underneath the Arctic surface waters, where it continues to flow at water depths of greater than 400 m along the contours of the Arctic Ocean. I suggest that this sinking and sliding underneath the surface currents between 24 and 18 ka occurred 1300 to 1400 km further south in the Norwegian Sea at approximately 66 to 67° North. But both the centre of this counterclockwise gyre may have shifted and its size may have varied.

At 17.7 ka, only a few hundred years after the Earth emerged from the deep chill of the last ice age, a major perturbation of the surface currents occurred when the Barents Ice Sheet disintegrated. This disintegration, and the large amounts of icebergs that it produced, changed the surface currents in the northern Norwegian Sea to primarily southward as opposed to the more westward direction encountered previously. Two large calving bays developed: one north of Norway, and another south of Spitsbergen. These areas were the parts of the ice shield where the most massive loss of ice occurred. Icebergs from north of Norway drifted to the Vøring-Plateau on a relatively straight path, and icebergs from south of Spitsbergen first drifted westwards near the ice margin and then curved towards the south. Two northward currents established themselves between these basically southward moving currents: one close to the Norwegian coast, and another one through the centre of the Norwegian Sea. These two currents evolved into the modern Norwegian Current–Norwegian Coastal Current system during the course of the next approximately 7000 years.

This evolution was probably not gradual but stepwise, with occasional far northward protrusions of the northward currents and movement in the opposite direction owing to icebergs from the BIS and adjacent ice sheets. Depending on the degree of mixing (lateral as well as vertical) of the northward moving, warmer, and higher saline waters with the southward-moving, colder, less saline, and iceberg-carrying waters, the northward current would alternate between a surface and a sub-surface flow. Only after the iceberg sources in the Pleistocene ice sheets were exhausted could the modern-day circulation fully establish itself.

The same southward-moving currents that brought icebergs from the BIS to the Vøring-Plateau during the Late Pleistocene also reappeared episodically during the Holocene. In this case, the current carried ice and perhaps also icebergs from places between the Svalbard archipelago and the East Siberian Sea. The ice drifted around the eastern side of the Svalbard archipelago across the Barents Sea and then straight towards the south through the Norwegian Sea. This reversal of the prevailing surface currents in the Norwegian Sea occurred frequently during the Holocene and was no isolated event. Thus it does not represent an insignificant aberration from the 'normal' climate situation (the one with the Norwegian Current), but rather reflects a very important, underlying second mode, or stade, of the Holocene climate over the Norwegian Greenland Sea region: apparently, the strong northward-moving Norwegian Current occasionally breaks down and allows cold waters with pack ice and icebergs from the eastern Arctic Ocean and the Barents Sea to enter the Norwegian Sea at least as far south as 66° N. The most important implication of this is that the modern Holocene climate is not stable. Instead it oscillates between two modes: one in which the Norwegian Current is strong and the climate is generally warm to temperate, and a second in which the Norwegian Current is weak and the climate becomes colder.

All samples of Holocene age in the Norwegian Sea were invariably affected by ice rafting from the Arctic Ocean, which itself is synonymous with a weakening of the Norwegian Current. As for the likelihood of this current reversal occurring within a specific time period, one can only make a rather vague estimate. It must have occurred within the time span that the samples represent. The shortest such period is 650 years, an approximate value due to uncertainties regarding the chronology and bioturbation. For most samples, the time spans are longer, usually between 1000 and 2000 years. Therefore the periodicity of the Arctic ice rafting episodes was less than 500–600 years. But because the IRD grains are sprinkled throughout the Holocene sediments (as seen in X-ray photographs) and not in distinct, thin layers separated by sediments without IRD grains, the occurrence was probably more frequent. We must also factor in that pack ice in the Arctic Ocean and the Barents Sea around Svalbard contains mostly fine-grained sediment in the clay to silt size range. If this had been the type of ice that drifted into the Norwegian Sea, it would have dropped clay and silt, but not the much coarser fragments that are the subject of this book. During the entire duration of the 1989 expedition with the research vessel 'Polarstern' into the Arctic Ocean, Barents and Greenland seas, only one quartz grain over 0.5 mm in size was found in sea ice sediment, despite the fact that a large swath around the ship's cruise track was consistently surveyed by helicopter every day for a seven week period, and that specifically ice floes with heavy sediment concentrations were sampled. All of this and much of the clean, sediment-free ice in the Arctic Ocean would not have left a trace in the >0.5 mm IRD record of the Norwegian Sea if it drifted there. Plus, as described earlier in the introduction, even icebergs which are commonly regarded as prime carriers of coarse particle loads can in some cases contain exclusively fine-grained sediment (consider, for example, the grounded iceberg in the Barents Sea with fine-grained eolian silt and sand. Its particle load, too, would not be noticed in the >0.5 mm record). For these reasons, most of the ice that drifted into the Norwegian Sea during the Holocene would not have left a trace in the coarse IRD record. Ice rafting from the Arctic Ocean into the Norwegian Sea was thus more frequent than once in 500 to 600 years. Most likely these events occurred on multi-decadal time scales.

In fact, these events could be the equivalent of the so-called 'great salinity anomalies' (Häkkinen, 1993), during which the melting of large amounts of ice in the Greenland and Iceland seas reduces their surface salinity and stops or impedes the formation of North Atlantic Deep Water (NADW), which in turn is one of several driving forces of the present-day surface currents with the temperate North Atlantic Drift. As a consequence of reduced NADW formation, the North Atlantic Current becomes weaker and very cold conditions take hold of Europe.

5.9.1 Possible driving forces

In my view, there are two possibilities for the causes of the current reversals/coolings in the Norwegian Sea during the Holocene. The first possibility is a chain reaction involving several steps, and the second is a more direct cooling mechanism.

The first possiblity is triggered by a change of the atmospheric circulation over the Arctic Ocean (step 1). The circulation must change so that the prevailing wind direction is from the Bering Strait across the North Pole towards Fram Strait. In this case, ice would not circulate within the Arctic Ocean (= no communication with the lower latitude waters), but would be pushed out of the Arctic through Fram Strait instead (= strong communication). The ice would continue floating southward into the Greenland and Iceland seas where it would melt in the summer (step 2). This melting would inject significant amounts of freshwater into the Greenland Sea, reducing its surface salinity (step 3). The reduced salinity would, at otherwise unchanged temperature, lower the density of the surface waters compared to times when there is less ice. When cooling of the surface waters resumes during the following winter, the colder winter waters and the waters resulting from brine rejection during sea ice formation would no longer be sufficiently dense to sink and contribute to NADW formation (step 4). As a consequence, the northward flow of warmer waters into the Norwegian Sea would decrease (step 5), and the colder northern, ice-bearing waters could spread across the Norwegian Sea (step 6). In this case, ice and cold surface waters could flow southward into the Norwegian Sea because the northward pushing and warming force of the Norwegian Current was diminished. An amplifying side-effect of the proposed atmospheric circulation over the Arctic would be that open water would form in the western part of the Arctic Ocean, in the Chukchi, Beaufort and East Siberian seas. If operating in the winter, these open leads would immediately refreeze, producing more ice, and this ice would be directed towards Fram Strait, adding considerable amounts of ice to be exported.

The second process causes a reduction of the strength of the Norwegian Current in a more direct way. The atmospheric circulation over the Arctic would have to be arranged so that winds blow preferably from 140–160° West (Alaska) to an area between 20° West and 60° East, and the air flow had to be wider than in the first scenario. In this case, ice would drift in a very broad swath through both the Fram Strait and through the Barents Sea. The difference to the first scenario is that ice would spread southward across the entire width of the Norwegian Greenland Sea, actively driven by the force of a strong atmospheric circulation over the Arctic. The necessary ingredients to drive both proposed processes is a zone of persistent high atmospheric pressure over both Siberia and northern North America. For the first scenario, the highest pressure had to be over the far eastern part of Siberia. For the second scenario, the area of maximum atmospheric pressure had to be over Alaska and northwestern continental Canada.

The reconstructed ice drift vectors are the result of both oceanic and atmospheric forces. The strongest identified atmospheric component is the distribution of high pressure to the north of 60° N. During the Pleistocene, high pressure cells existed over the large ice sheets. For the Norwegian Greenland seas, the most important high pressure areas were located over the combined Fennoscandian (FIS) and Barents/Svalbard (BIS) ice sheets and over the Greenland Ice Sheet (GIS). Of those three ice sheets, the influence from the BIS was strongest in the northern Norwegian Sea. During the Holocene, a belt of high pressure occasionally developed over eastern Siberia and northwestern North America between 60 and 73° N.

The oceanic component of the circulation system can be thought of as a large closed loop in which waters, ice and icebergs from the Norwegian Greenland Iceland seas communicate with the waters of the Arctic Ocean, the North Atlantic and the Barents Kara seas (only during times when the latter two were not glaciated). The loop is a counterclockwise circulation in which waters move northwards in the east and southwards in the west. Waters, sea ice and icebergs that exit the Greenland Iceland Sea to the south mix with northward-moving North Atlantic waters and re-enter the Norwegian Sea through the southeast. The loop is closed either in the Norwegian Sea (Weichselian Maximum), in the Fram Strait (today), or at some location in between. In the following sections on ice rafting in the Arctic Ocean it will be demonstrated that this loop at times extended deep into the central, eastern and even western Arctic Ocean.

5.10 PRE-WEICHSELIAN GLACIATIONS IN THE NORWEGIAN GREENLAND SEA

The IRD from glaciations older than the Weichselian is preserved in cores from the western Norwegian Sea. There, the lower sedimentation rates have led to a longer record at the expense of high stratigraphic resolution. The long kastenlot cores 23059 23060, and 23062 have nearly identical sediment sequences reaching back to approximately 350,000 years bp. The example presented here is core GKG-KAL 23059. This core has a position almost exactly in the centre of the NGS. Its closest distance to any of the surrounding landmasses of Greenland, Iceland, Norway and Spitsbergen is circa 750 km. Owing to its geographic position, this core could have theoretically received IRD from all of the aforementioned sources. But the IRD composition (Fig. 5.31) is dominated by clastic sedimentary rocks from isotope stages 12 through 6. During this period, clastic rocks make up at least 50% of the

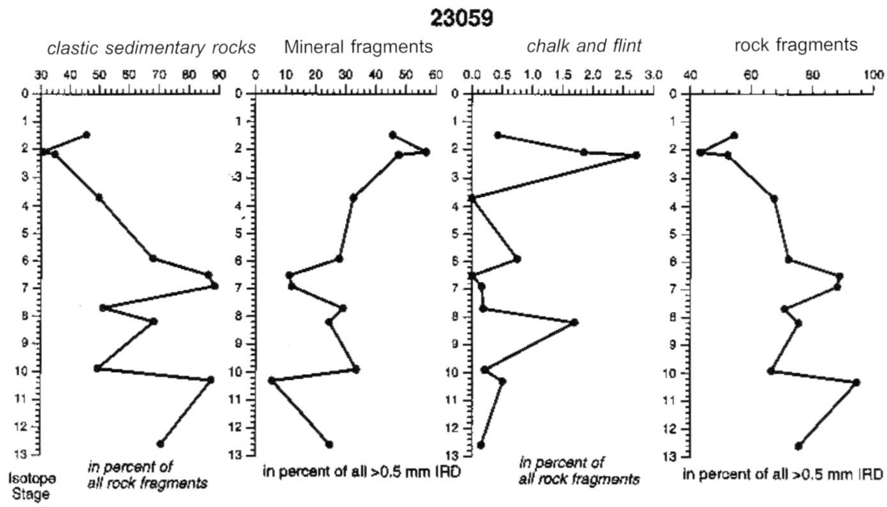

Fig. 5.31. Composition of >0.5 mm IRD in core KAL 23059, western Norwegian Sea. Note that the IRD data are presented as percents of all >0.5 mm grains (right and second from the left) and in percent of all rock fragments (left and second from the right). The core is located at 70.18.3 North and 3.07.3 West, water depth: 2283 mbsl. The plots show the combined GKG and KAL records. The vertical scale is in isotope stages. Every sample was given a decimal number that denotes their stratigraphic position within an isotope stage. For example, a number of 2.1 would be a sample taken from the top of isotope stage 2, a number of 5.9 would be from the bottom of isotope stage 5, and so on. The lowest sample is from the core catcher (600–650 cm core depth) and thus from a deeper stratigraphic position than those samples for which isotope values are available. It was assigned an isotope stage 12 age from 425,000–450,000 years before the present. KAL 23059 contains the oldest IRD record from the Norwegian Sea and contains debris from four pre-Weichselian glaciations, isotope stages 6, 8, 10 and 12.

IRD rock fragments, and can be as abundant as 90% of the rock fragments (late stage 10 and early stage 6). After stage 6, the proportion of clastic sedimentary rocks among the IRD drops and gives way to increasing amounts of crystalline rocks and feldspar grains.

Keeping in mind that (a) these samples were preferably taken from some of the coarsest units in the core, (b) that the samples represent up to 50 cm thick core intervals, and that (c) the samples are spaced 30 to 60 cm apart, the data shown in Fig. 5.31 are composites that integrate across 20 to 50 cm of sediment thickness. The IRD composition between the sampled intervals is still unknown.

It nevertheless appears that the lower part of the core, stage 6 and older, is profoundly different from the Weichselian Glaciation. Clastic sedimentary rocks dominate among the rock fragments prior to circa 120 ka. On average, they constitute 70% of the ice transported rock fragments, at a variability of +/− 20%. This, the petrographic similarity of these clastics with those from the Barents shelf, and the fact that samples from this core were placed into the same cluster as samples from directly off the Barents shelf (see Table 5.4, cluster 7) demonstrates that the Barents shelf–Svalbard area was the most important source of IRD between 370 and 120 ka. The presence of chalk and flint in the same samples (Fig. 5.31), however, indicates that at least some mixing with ice rafted lithologies from the southeastern entrance of the Norwegian Sea occurred. The samples therefore are not pure, entirely derived from Svalbard and the Barents shelf, but contain a certain amount of advected components from sources as far away as southern Scandinavia.

But the important point is that the change towards a more Norwegian-dominated crystalline type of IRD composition does not occur prior to the last, the Weichselian, glaciation. And this despite the fact that the pre-Weichselian glaciations in the terrestrial records from northern Europe were all more extensive than the Weichselian Glaciation. During each of the three known precursors, the Saalian ice age with its two stages, the Warthe and Drenthe stade, and the even older Elsterian ice age (equivalents to marine isotope stages 6, 8 and 10), the margin of the Fennoscandian Ice Sheet extended several hundreds of kilometres farther to the south than during the last glaciation (Woldstedt, 1965; Woldstedt and Duphorn, 1974). One would therefore expect, at least in theory, that the Norwegian-derived crystalline rock IRD should have been dispersed deeply into the Norwegian Greenland Sea. Instead, these older glacial stages were dominated by IRD of northern, more sedimentary, provenance, and the direction of the surface currents was predominantly towards the south. This is also further corroborated by the occurrence of coal from Siberia in sediments of stage 6 age in the central NGS (Bischof et al., 1990).[23] Thus, for unknown reasons, the driving forces of southward movement of ice and surface waters were stronger relative to the northward moving forces during these older glacial stages. The southward currents moved through the geographic centre of the NGS (Bischof et al., 1990; Fig. 5.3.2), which amounts to a complete reversal of the flow pattern shown in Fig. 5.16b. Whether the driving forces were either strong winds over the Arctic domain blowing to the south or a sub-surface inflow of North Atlantic waters at a deeper level through the southeastern Norwegian Sea, cannot be answered.

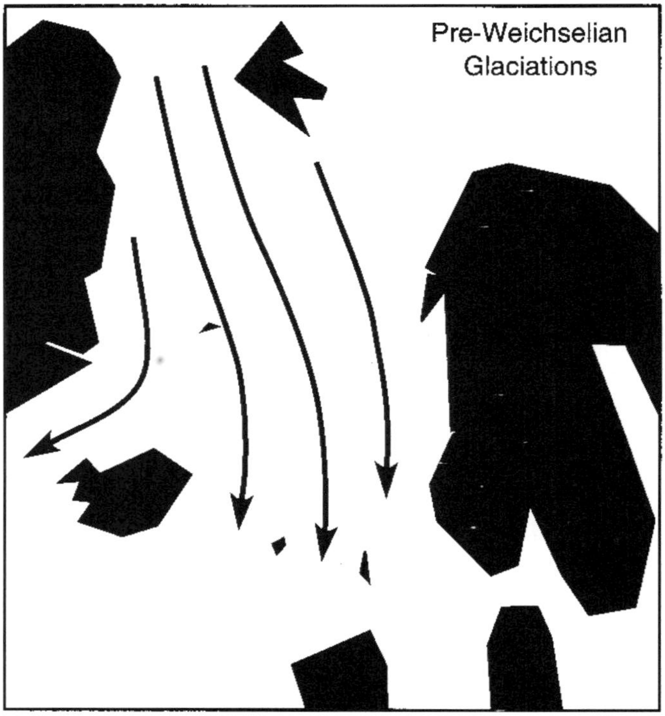

Fig. 5.32. Generalised ice drift during Pre-Weichselian Glaciations. This is the drift direction that ice and icebergs probably took during oxygen isotope stages 6, 8, 10 and 12. The proportion of Scandinavian-derived crystalline IRD was reduced by abundant icebergs from sedimentary source terrains in Svalbard and the adjoining Barents Sea, and places as far away as Siberia.

5.10.1 Pre-Weichselian terminations

There is another sedimentological aspect of the pre-Weichselian glaciations that deserves attention: as we have seen in Section 5.1 on Norwegian Sea glaciomarine sediments, the glacial cycles before the Late Weichselian Glaciation (marine isotope stages 6, 8, 10 and 12) terminate with dark grey, dark greenish-grey, or even black, sandy to gravelly mud. This sediment is firm to very firm, significantly more consolidated than sediment above and below, and contains ice rafted coarse grains in maximum concentrations. These dark layers also have a relatively high carbon content, and the carbon is of terrestrial origin (Henrich et al., 1989). The IRD of these distinct layers is very rich in clastic sedimentary rocks with relatively high proportions of black shales and some coal. In isotope stage 6, up to 21% coal fragments were observed in IRD of the central NGS (Bischof et al., 1990). The dark layers rich in clastic sedimentary rock IRD in the central NGS are much thicker (30 cm or more) than the Late Weichselian equivalent that resulted from the disintegration of the Barents Ice Sheet (only 4–6 cm on the northern Vøring-Plateau).

If I may take the freedom to speculate about the causes that produced these peculiar terminations, here is a possible (theoretical) scenario. If the strong southward currents with icebergs from the north (Svalbard–Barents shelf region, plus, in case of stage 6, Siberia) received their kinetic energy from northward-flowing deeper waters that entered the NGS from the southeast, the deepest-keeled icebergs could have been in contact with somewhat warmer waters, and this would result in an increasing drop-out rate of the basal debris which is mostly concentrated at the base of the parent ice sheet. This would explain the comparatively great thickness of the dark IRD layers. Since the coarse IRD grains are accompanied by finer grained clastic debris of essentially the same petrographical and geochemical composition, and since further much of the IRD consists of dark, carbon-rich clastic rocks, a large reservoir for terrestrial carbon existed (which explains the high carbon content and its terrestrial origin). This carbon could react with and consume the free oxygen at the sea floor and thus create nearly anoxic conditions. And, further, since most black shale formations (from where the black shales in the IRD were derived) contain elements such as iron, copper, lead, manganese, sulphur, selenium, etc., these elements could react and precipitate as microscopic crystals in intergranular space, displacing pore water, and cause the overconsolidation of the dark layers. The same reaction could also explain why overconsolidated sediments of a presumed stage 6 age exist on the Lomonosov Ridge (Jakobsson et al., 1998). Instead of overconsolidation by chemical mineral precipitation in pore space, these authors suggest that the overconsolidation was the result of a grounded glacier, even though this explanation is followed by a question mark. The glacier theory could be problematic from a physical point of view, since it requires grounded ice in more than 1000 m of water. But nowadays, the thickness of free floating ice shelves such as the huge Antarctic ones is limited to 400–500 m. If this thickness represents the maximum thickness that a floating ice shelf can attain because of the ice rheology, the grounded glacier theory must be rejected and another mechanism found to explain the overconsolidation.

But this author is not a geochemist and the proposed reaction chain should be understood as my contribution to the debate. In my view, however, it is more than mere coincidence that an extreme state of the NGS surface circulation, predominantly southward flowing, iceberg-laden waters in the central NGS, occurred during the same time period during which an equally unique type of sediment developed. And therefore, the detected changes of the surface circulation may have affected the bottom water circulation and the chemical conditions at the water-sediment boundary. In other words, the observed circulation changes were 3-dimensional, not confined to the surface waters.

6

Ice rafting in the Arctic Ocean

6.1 INTRODUCTION

The Arctic Ocean is a fascinating study object from a climatological perspective because of its permanent ice cover. Compared to its vast aerial extent of approximately 13 million square kilometres (the Arctic Basin *sensu stricto* plus its adjacent shelf seas, the Chukchi, East Siberian, Laptev, Kara and Barents seas), the ice cover is incredibly thin, only 3 m on average. It also has, in relation to geological time scales, a fairly brief history. According to Clark et al. (1980) and Clark (1990), the Arctic ice cover must have existed at least since late Pliocene time because of the accumulations of glacial marine sediments since then. The ice cover is seasonally variable: while all of the Arctic is frozen in the winter with air temperatures as low as $-60°C$, in the summer almost half of the Arctic ice cover melts and disappears. Because of this extreme seasonality, the ice cover is thought to be very sensitive to climate, in particular temperature changes. The basic idea is that global warming, if it is a reality, should immediately lead to a thinning and shrinking of the ice cover. And indeed, the SHEBA experiment (explained in endnote 3) ended up with a net loss of 10 cm ice thickness after one year, and more recently Kwok and Rothrock (1999) demonstrated that the regional extent of the Arctic ice cover also seems to be shrinking, based on satellite measurements of the size of the Arctic ice cover. But whether these observations indicate a significant response to climate change, as opposed to naturally-occurring climate changes, remains unclear.

The Arctic Ocean and its ice cover play a very crucial role in the global temperature balance, for multiple reasons. At the present the Arctic Ocean is a heat sink, that is an area in which the Earth loses heat. The major gateway between Arctic and sub-Arctic seas is the Fram Strait. In addition to this, other marine connections exist in the Barents Sea, in Nares Strait (a narrow channel between Ellesmere Island and northwest Greenland), the Canadian Arctic Archipelago (which is quickly developing into an area of intense scientific interest), and through the Bering Strait. The incoming, North Atlantic waters enter the Arctic Ocean through the eastern

Fram Strait at a rate of approximately 3 Sverdrup (1 Sv = 1 million cubic metres of water per second) and a temperature of 2–4° C. But the outflowing cold Arctic waters of the East Greenland Current have a temperature between zero degrees and the freezing point of seawater, approximately −1.8° C. Because the East Greenland Current also flows at the same rate of 3 Sv, the (water) mass balance is even but the temperature balance is negative, so the heat of the ocean is lost to the atmosphere. At least some of this loss is due to sea ice formation. When the Arctic seawater starts to freeze in early winter, the phase transition from liquid water to solid ice consumes heat which comes from the surface waters.

Another element of climate sensitivity is the reflection of incoming solar radiation from the glistening white ice surface of the Arctic Ocean. The temperature at the surface of the Earth is the product of the ambient air temperature and the conversion of the incoming solar long-wave radiation into heat. The latter depends on the type of surface material hit by the solar radiation. If the surface is dark, much of the radiative energy is converted into heat, but when the surface is white and highly reflective such with as ice and snow, then much of the radiation is simply reflected back to the atmosphere without a temperature effect (see endnote 7 on albedo). This means that even during the summer months of highest solar insolation, the temperature at the surface of the Arctic ice cover rarely rises above zero. Thus, the permanent ice cover stabilises cold air temperatures and maintains its own existence. At the present, the balance between the volume of ice that melts in the summer and the volume that refreezes in the winter is approximately equal.

But when the albedo (the amount of reflected solar radiation) is greatly reduced, the summer melting could increase, melting more ice. And with less ice and increasing areas of open water, the albedo would drop further and amplify the warming trend. This warming feedback loop could eventually remove the Arctic ice cover. Decreasing albedo can be caused by melt ponds which develop in the summer on the ice surface, sediment inclusions, and organic matter (in the Arctic primarily diatoms and brown algae). Decreasing albedo, of course, could also result from extreme northward intrusions of temperate Atlantic waters which could melt much of the ice through contact with the warmer waters. As we have seen in Section 5.4.4, such a scenario existed briefly during the middle Holocene but was not strong enough to melt the Arctic ice cover.

Equally, if not more important, for the climatic development in the Arctic region than the annual freeze and thaw cycles of the ice cover are the pack ice drift direction and the surface ocean circulation. When the Arctic communicates strongly with the sub-Arctic seas by exporting ice into them, the area of ice-covered, white, highly reflective surface in the northern hemisphere expands, resulting in a cooling effect through increasing albedo and reflection of the incoming solar radiation. But when the Arctic is not communicating, for instance when ice drifts in a closed loop resolved within the Arctic Ocean, then the northern hemispheric albedo decreases and the thermal balance changes towards warmer temperatures.

The pack ice movement and surface circulation nowadays is somewhat of a hybrid between the two aforementioned states. It consists of the Transpolar Drift, a strong current that moves ice from the eastern Arctic Ocean across the Barents Sea and through Fram Strait into the NGS. That is the communicative part of the Arctic circulation. The other element is the Beaufort Gyre, a clockwise circulation in the western Arctic Ocean in which ice can remain trapped for decades. For example, ice island T-3 spent more than three decades in the Beaufort Gyre before exiting the Arctic Ocean through Fram Strait. The Beaufort Gyre would be the non-communicative part of the Arctic circulation. The depiction of a closed loop in the Beaufort Gyre and the comparatively straight Transpolar Drift is shown in virtually every paper ever written on the Arctic Ocean circulation. Therefore I like to call this the conventional view of the Arctic surface circulation. However, in a preview of the next chapters, I want to mention here that the geologic (ice rafted) record does not support the concept of the Beaufort Gyre, and that the true long-term average ice drift as well as its shorter-term variants deviate considerably from the conventional view of the Arctic circulation.

The present-day ice drift is highly variable and changes incessantly between communicative and non-communicative states. The monthly averaged ice drift vectors can be viewed at the web site of the International Arctic Buoy Program which is maintained by Ignatius Rigor and Mark Ortmeyer at the Applied Physics Laboratory of the University of Washington (http://iabp.apl.washington.edu/Summary/). I strongly recommend that all those interested in climate change visit this web site, because the ice drift vectors are based on actual, satellite-based observations of drifting buoys whose positions are continuously monitored. Therefore, the ice drift vectors shown on this web site are the most realistic pictures of the Arctic ice circulation ever obtained. The ice drift directions vary strongly on a monthly basis, and one can clearly see that they are not stable. The University of Washington web site also shows the surface air pressure and temperature and a composite of the ice drift vectors and surface air pressure. On that composite, one can clearly see that the ice drift responds directly to the prevailing wind direction. In other words, the Arctic pack ice drift is controlled by the atmospheric circulation.

A direct response of the drift directions of relatively thin (3 m on average) pack ice to the atmospheric circulation is in agreement with the conclusions from the ice drift patterns in the NGS, which in the northern part are more atmospherically controlled whereas in the southern part oceanic forces dominate. In case of the NGS ice drift, the large Pleistocene ice sheets over Scandinavia, the Barents Sea–Svalbard region, and Greenland created high pressure cells whose air flow drove the ice drift and surface circulation. In fact, the Pleistocene ice drift direction of a large counter-clockwise rotation in the Norwegian Greenland Sea can be explained with wind forcing from the two large atmospheric high pressure cells on both sides of the NGS alone and does not necessarily require the surface ocean as a driving force. Similarly, the present-day ice drift in the Arctic Ocean is driven mainly by a quasi-permanent area of high atmospheric pressure over Siberia.

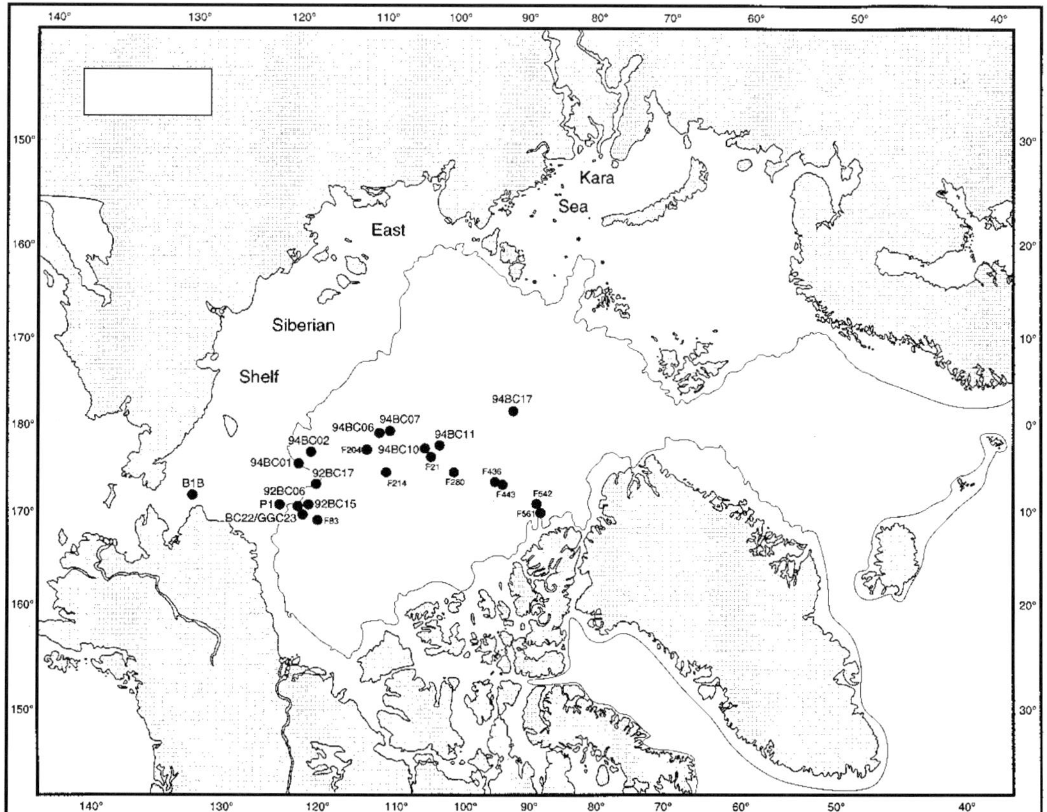

Fig. 6.1. Map of all Arctic Ocean core locations analysed for IRD composition thus far. In addition to those cores analysed for the very coarse >250 μm IRD composition, there are several more (94BC01, -02, -06 and -07) that have thus far only been analysed for the geochemical composition of individual iron oxide minerals in the size range 45–250 μm. These cores were not analysed for >250 μm clast composition, because inspection of the content of the sample containers showed that their IRD petrographies are identical to those of the neighbouring cores. Core P1 is among the group at the lower left.

6.2 ARCTIC OCEAN IRD

In this section, the most important characteristics of Arctic Ocean IRD will be presented, and the IRD composition will be related to circum-Arctic bedrock formations in order to trace the IRD back to its sources. The individual pieces of evidence will then be blended into a model of the Arctic Ocean circulation and the climate changes that this system has undergone during the last 1.5 million years. The information presented here is based on quantitative analyses of the petrographic and mineralogic composition of >250 μm IRD in 11 piston and six box cores from the Arctic Ocean (Fig. 6.1). We (Dennis Darby, David Clark and I) have also acquired over 300 source area samples from the periphery of the Arctic Ocean (Fig. 6.2). These samples are mostly shallow marine surface sediment samples which represent the petrographic composition of their respective hinterland.[24] They relieve us of the need of making educated guesses about what type of IRD can be expected from certain parts of the circum-Arctic. But even without having these source samples, most of the following conclusions would still be possible, simply because the composition of some types of IRD is so extreme that the composition alone limits the possibilities to just one. For example, the carbonate percentages in many IRD samples are so high that only Victoria Island and nearby islands in the southwestern Canadian Arctic Archipelago could have produced them. There is simply no other area with similarly carbonate-dominated bedrock and comparable geographic extent. By the same line of reasoning, the fact that extremely high abundances of crystalline rocks can be found in two cores from close to the northern coast of Ellesmere Island but nowhere else can only be explained with a derivation from North Ellesmere Island.

The foremost characteristic of coarse Arctic Ocean IRD is that it is almost invariably present. Only some piston cores from the Northwind Ridge (the area to the centre left in Fig. 6.1) and a few box cores from the Wrangel Abyssal Plain (94BC10 and -11) contain sediment sections with finely-laminated clay and silt but no coarser ice rafted fragments. The relative amount of coarse IRD in the cores has changed greatly through time. During the last approximately 1.45 million years, 13 coarse IRD peaks are counted in core Fl-561 from close to the coast of Ellesmere Island (Fig. 6.3). The rather regular spacing between the peaks suggests a rhythmical sequence with peaks every 111,000 years, approximately the cycle of major global glaciations. Therefore the coarse units are thought to be glacials during which large continental and partly marine-based ice sheets existed in both polar regions on Earth. These ice sheets terminated in the sea and ejected large amounts of icebergs which transported, and lost, terrigenous debris from bedrock around their parent glaciers throughout the ocean. This process led to an increase in the relative proportion of coarse IRD in the deep-sea sediments of the Arctic Ocean. Conversely, after the end of the glacial periods, the iceberg sources no longer existed and ice rafting continued at comparatively low intensity. During these interglacials, ice rafting was accomplished mostly by sea ice and only few icebergs. Consequently, the amounts of coarse IRD during these intervals are low. The differences between the interglacial, low levels of IRD deposition and those of the Pleistocene glaciations are

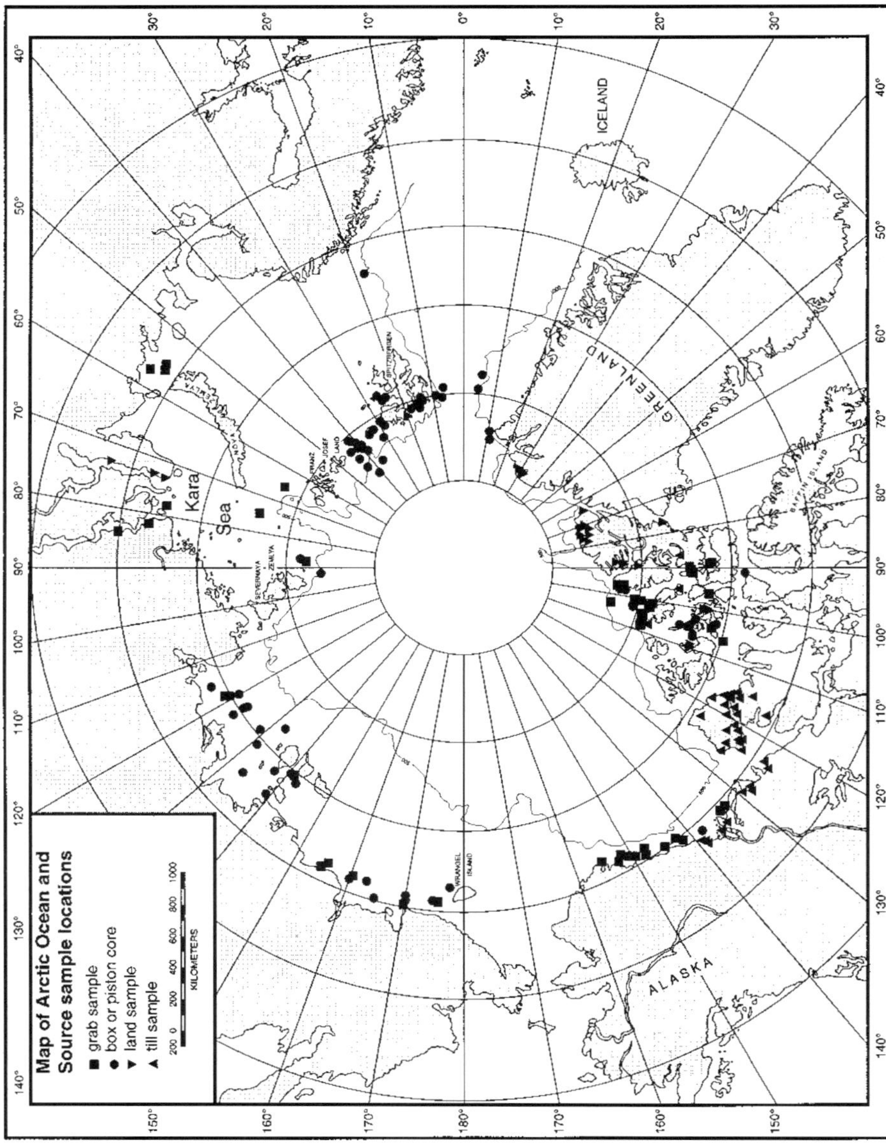

Fig. 6.2. Map of all source area samples to which our core samples have been compared.

drastic: in core Fl 561 the values fluctuate between fractions of a percent and as much as 12 weight percent >250 μm (Fig. 6.3). But this 3 order of magnitude difference is not the full range of variability in the Arctic Ocean, because in core Fl-561, 4 cm-thick sections of sediment were sampled. Therefore, the weight percentages shown in Fig. 6.3 are composites that integrate across the sampled interval. If, for example, a thin, fine-grained layer existed in an otherwise coarser sediment section, it would only result in an unnoticeable drop of the IRD weight percentages, but the true fineness of the layer would be missed. As part of an ongoing research project, the coarse IRD content in a sediment core from the western Arctic Ocean (piston core P1) was measured as IRD grains >250 μm in one gram of sediment. The values oscillate between 0 and 10 grains/g sediment in the Holocene (younger than 9000 years bp) section (Fig. 6.3a), and between 100 and 300 grains/g in the Late Pleistocene section (Fig. 6.3b). This amounts to 4 orders of magnitude difference, with even the highest Holocene values still an order of magnitude below the lowest Pleistocene values. These drastic differences confirm that the frequency of icebergs,

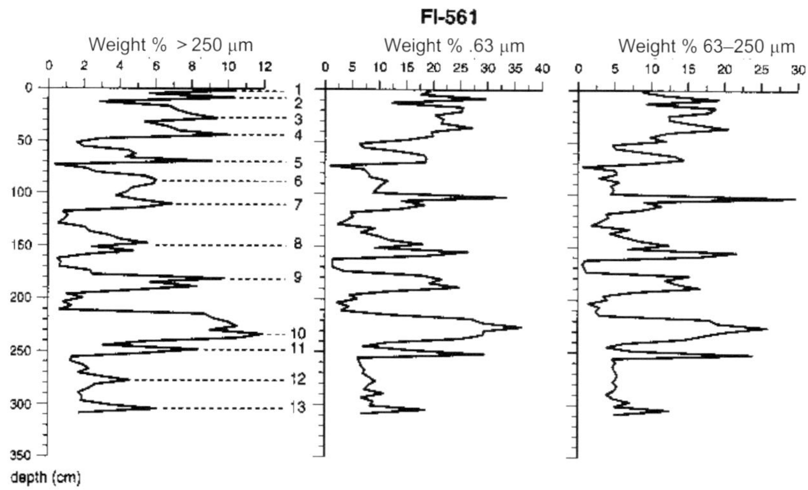

Fig. 6.3. Textural characteristics of an Arctic sediment core (Fl-561) from just north of Ellesmere Island. The left side shows the weight percentages of the >250 μm fraction which was analysed for IRD composition. There are 13 IRD peaks within the time span represented by this core, whose bottom is Arctic sediment unit F (Clark et al., 1980) with an age of 1.4–1.5 Ma (Bischof et al., 1996). The spacing of the peaks is rather even, which further suggests that the time intervals which separate them were regular. Thirteen peaks within 1.45 Ma give an average time interval of circa 111,000 years, very close to the time intervals between major global glaciations. Therefore, each of these peaks is considered a major glaciation (= even numbered marine oxygen isotope stages) during which massive amounts of icebergs drifted across the core site. The shape of the coarse fraction curve is almost the same for the 63–250 μm (fine sand) fraction, right side, and for the composite, the total >63 μm fraction. Percentages of up to 12% for the >250 μm fraction, and values of over 35% for the >63 μm fraction are among the highest from anywhere in the Arctic Ocean.

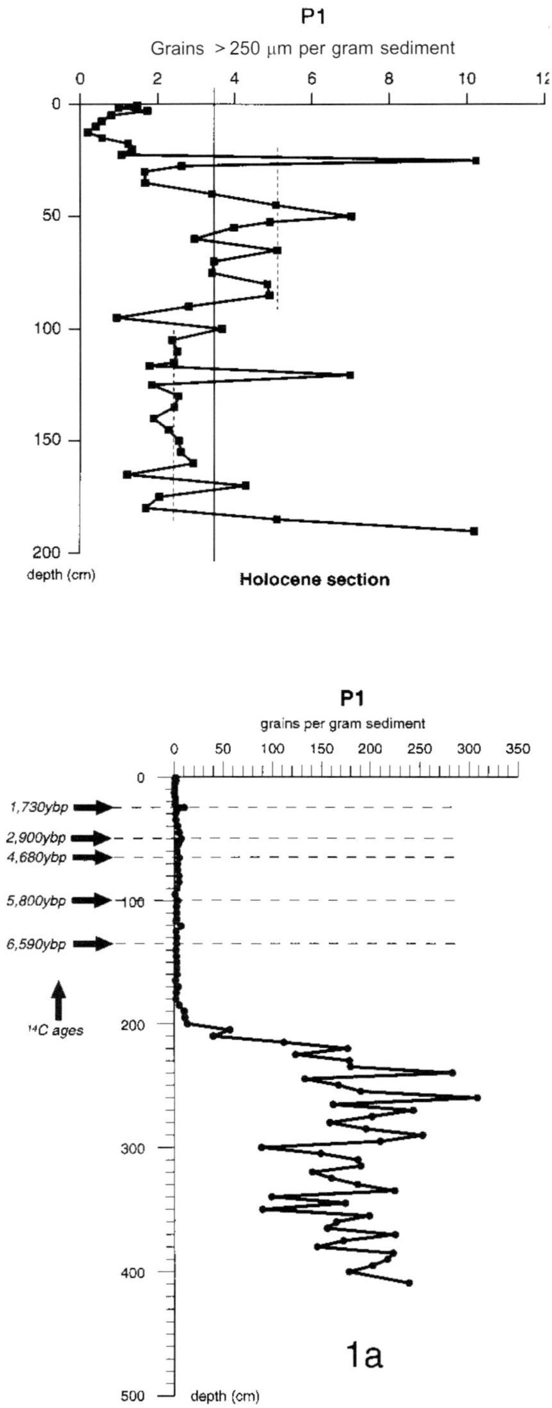

(a) IRD density in core P1, expressed as IRD grains >250 μm in one gram of sediment. This is the Holocene section of the core with a variability between zero and slightly over 10 grains/g sediment. The Holocene grain density fluctuations thus are within 3 orders of magnitude. The vertical line denotes the Holocene average. Note that the grain densities during the early Holocene (broken line on lower left) are lower than those during the mid to late Holocene (broken line on upper right). Once again, a change in the style of ice rafting is observed in about the same stratigraphic horizon as in the Norwegian Sea.

(b) Same as in (a), but for the entire length of the core. This graphic contains the Pleistocene section as well. Compared to the high grain densities and their dramatic oscillations in the Pleistocene, the Holocene variability disappears and looks almost like a flat line.

and not other factors, have the most immediate impact on the deposition rates of coarse-grained IRD. These other factors are the concentration of sediment in the ice and icebergs, the surface water temperature, and the intensity of summer melting. The highest IRD grain densities recorded thus far in the Arctic Ocean are from another core from the Northwind Ridge area, box core 92 BC 17: 1660 IRD grains per gram sediment. Therefore the full range of IRD grain densities is 5 orders of magnitude.

The reason that the grain frequencies of P1 were plotted in two separate diagrams, one for the Holocene and another for the whole core, is to emphasise the importance of a sufficiently long paleo-record for climate research. By far the majority of climate change-oriented articles published at the time this book was written rely on observations made over the course of several years to a few decades. This is so because reliable measurements of climate parameters such as temperature or pack ice thickness are only available for the last 20 to 30 years, with the earliest such measurements going back 150 years. While certain tendencies within this time range may be found, it is totally unknown how these changes compare to the underlying, naturally-occurring oscillations of the climate over longer time spans. A good example is that of temperature records which show a slight increase during the last 150 years, a fact that is often seen as an indication that global warming is under way. Another example is shown by more recent publications which show a thinning and shrinking of the Arctic ice cover, apparently (at least it is tempting to think) in response to global warming.

But how do these trends compare to the greater picture of the longer time series, for example the entire Holocene? Figure 6.3a gives a partial answer. The average Holocene sedimentation rate of this core is 25 cm/kyr (see radiocarbon dates in Fig. 6.3b). Since the samples analysed are 2 cm thick, on average, they each represent 80 years. Therefore, the observational time period for which atmospheric measurements are available (150 years) equals the two top data points. The sediment property shown in Fig. 6.3a is partly the result of sedimentologic processes, but also to a great extent climatically influenced. For the IRD grains to be dropped at the site of P1, they first have to get there (= ice drift direction), and then they have to be released from the transporting ice, which is best accomplished by melting. Thus the IRD density record of the Holocene at this site may very well be a record of the intensity of summer melting, thus an indirect (summer) temperature record. But be this as it may, we can consider the IRD record the equivalent of the record of a (or any) climate variable. It could be one or a combination of several climate parameters such as surface water temperature, air temperature, sediment concentration in the ice, ice drift velocity, intensity of summer insolation, sea ice extent, and so on. In any case, Fig. 6.3a shows the variability of this (these) parameter(s) throughout the Holocene.

We register that the variability at the top, the equivalent of the 150 years for which climate observations are available, is very small compared to the variability during the Holocene which, in turn, is dwarfed by the huge variations in the Pleistocene section. Therefore, if we consider this the record of past summer surface temperatures, then the variability recorded during the past 150 years by man-made measurements is negligible compared to the variability during the Holocene, and therefore unrepresentative for the Holocene climate as a whole. And because of this, high resolution marine sediment records that reach much further back are essential for a complete understanding of the global climate system.

From this standpoint it makes no difference whether the record of Fig. 6.3a is a reflection of the summer surface temperature alone or of a combination of several of the processes mentioned above, because they all are directly and indirectly climate related. Let us consider the volume of sediment contained in the pack ice. The amount of sea ice sediment fluctuates greatly from year to year (see Chapter 4 on the process of ice rafting and particle incorporation), which certainly affects the amount of sediment and coarse IRD that can be dropped in any given year. This sounds like a purely sedimentary, randomly-occurring process, but climate plays a very important role. For example, if winter conditions are cold and quiet (absence of storms) and the circum-Arctic lands are buried under a thick snow cover, the potential for particle entrainment in or onto the pack ice is minimal. In this scenario, the opportunity for particle entrainment only exists briefly during the spring when river runoff can spill fluvial sediments onto the still frozen ice pack. But when the winters are drier (less snow cover) and more dynamic (windy to stormy), sand grains can be blown from exposed land surfaces onto the pack ice, and agitation of the surface ocean can resuspend particles from the sea floor at shallow depths which subsequently become frozen into the newly forming ice pack. These processes would greatly increase the volume of sediment in the pack ice and thus, after some drift distance and melting in the summer, the amount of IRD in the sediment. Therefore, the process of sediment entrainment is clearly not random, but itself a reflection of changing climatic conditions.

Figure 6.3b shows impressively how the IRD grain frequencies reflect the vanishing of the Laurentide Ice Sheet. According to the terrestrial stratigraphy, the LIS retreated on land at approximately 9 ka. After this time period, Pleistocene iceberg rafting was over. The core reflects this beautifully with a sharp drop of the IRD grain frequencies to values near zero. Also, the carbonate shells of microscopic marine organisms such as foraminifera reappear after the IRD grain frequencies drop. And therefore, by analogy, sediments with similarly low grain concentrations but high microorganism content as in the upper 180 cm in core P1 generally reflect interglacial conditions with few icebergs, limited continental ice cover, and a sea ice cover similar in style to the one today. Intra-Holocene grain abundance variations are the result of changing summer temperatures, which in turn affect the thickness and geographic extent of the Arctic pack ice. Relatively high (measured by Holocene standards) IRD densities indicate higher summer temperatures and less ice cover, conditions under which more of the lithogenic debris in the ice would fall out.

And times of low abundances, again measured by Holocene standards, indicate quieter winter conditions and/or less summer melting, which would result in less particle fall-out.

During the Pleistocene, however, when IRD grain densities were several orders of magnitude higher than during the Holocene, the process that caused the IRD grain density fluctuations was different. Here, the fluctuations reflect iceberg discharges from different parts of the Pleistocene ice sheets under generally unchanged, cold temperatures, similar to the analogy of core 23074 in the Norwegian Sea, which had several distinct pulses of IRD during the Last Glacial Maximum, a period of flat, unchanged delta O-18 values. After this treatment of textural characteristics of Arctic Ocean sediments, let us now turn our attention to the composition of the IRD.

6.2.1 Arctic Ocean IRD composition

The IRD composition in deep sea sediments of the Arctic Ocean has been described in Bischof et al. (1996) and Bischof and Darby (1997). The most important differences to IRD in the Norwegian Sea are the frequent occurrences of carbonates in high percentages, the near complete lack of feldspar crystals, and the dominance of sedimentary over crystalline rocks. The Arctic Ocean IRD consists of three basic types: (1) carbonate-dominated, (2) quartz-dominated, and (3) crystalline rock-dominated. Type 1 (carbonate-dominated) IRD is by far the most important: it consists of up to 52% carbonate clasts in the >250 μm IRD, plus clastic sedimentary and crystalline rocks and quartz in lesser percentages. The carbonates are (in hand specimen) uniform medium brown to medium grey clasts without any diagnostic features. They are microcrystalline and unfossiliferous. Under the microscope, the dominant colours are white and medium brown, followed by (in decreasing order) grey, yellow, black, orange, red and green. This type of clast composition dominates in the central and western Arctic Ocean, but not so much in the vicinity of Ellesmere Island and North Greenland. The origin of these carbonates is Victoria Island with adjacent islands of the southwestern Canadian Arctic archipelago. Because of the occurrence of crystalline rocks, which are mostly white, clear, orange and red, the source area of the icebergs that transported type 1 IRD must have extended onto the Precambrian Canadian Shield. In other words, the origin of type 1 IRD coincides with the area occupied and covered by the Late Pleistocene Laurentide Ice Sheet.

In the central Arctic Ocean, type 1 (carbonate dominated) IRD alternates with quartz dominated IRD, type 2. The quartz-dominated samples appear as almost pure, clear quartz sand with only small amounts of non-quartz lithologies. The quartz percentages in type 2 IRD can be as high as 91%. Of the source areas around the Arctic Ocean, only four regions could have produced quartz in such high percentages:

(1) the western Queen Elizabeth Islands (QEI: the group of islands to the north of Victoria and Banks islands) where the bedrock consists of poorly consolidated Mesozoic clastics and Cenozoic quartz sands;
(2) the Laptev Sea, a shallow shelf in northern Siberia that receives debris from sedimentary and volcanic formations;
(3) the Kara Sea, another Eurasian Arctic shelf with abundant quartz sand; and
(4) the large estuaries of the rivers Ob and Yenisey in northern Russia.

Selecting the most likely source out of this group of four candidates is not easy. The quartz grains in quartz-dominated, very coarse Arctic Ocean IRD are mostly clean, devoid of any stains. But the quartz grains in the Ob and Yenisey estuaries and in bottom sediments of the Kara Sea are for the most part heavily stained, covered with strongly adhesive yellow, orange, and red iron oxide–hydroxide stains which give the samples a somewhat 'dirty' appearance. This would leave us with two potential candidates, the western QEI and the Laptev Sea. The Laptev Sea is considered never to have been glaciated (Kassens et al., 1995), but the quartz-dominated IRD almost always occurs in very coarse and abiotic glacial sediments, so the source of the quartz IRD must have been in an iceberg-productive, glaciated region. This would eliminate the Laptev Sea, and the western QEI, the location of the Innuitian Ice Sheet (IIS, Fulton, 1989; Hodgson, 1989) would have to be the source. But the following section will show that the case is more complex than this simple, approximate solution.

Type 3 IRD is dominated by crystalline rocks in percentages of up to 70% of the total >250 μm IRD. This type of IRD is geographically restricted to a small part of the Arctic Ocean north of Ellesmere Island, the site of core Fl-561 (Fig. 6.4). The crystalline rocks in this area (the site of Fl-561, Bischof et al., 1996 and its companion core Fl-542, Bischof and Darby, 1997) are in several aspects different from crystalline rocks from other parts of the Arctic Ocean: (a) they are much more abundant than elsewhere, (b) they contain half dark coloured and half light coloured crystallines as opposed to elsewhere where light coloured rocks are most abundant, (c) they contain meta-argillites in higher percentages than elsewhere, and (d) they contain a peculiar, very eye-catching metapelite with large amphibole blades in random orientation. This rock occurs in many different, radiant colours of white, silver, green, purple, grey and even blue. This type of metapelite was described in an unpublished technical report by Bjørnerud and comes from Yelverton Inlet, northwestern Ellesmere Island. It is a truly unique type of rock and was not found once in any of the other cores from the Arctic Ocean – and this despite the fact that I did not just count the grains in a random split, but also scanned routinely across the rest of every sample to check whether this particular rock type was present. Therefore the crystallines and the multicoloured metapelites have an origin in northern Ellesmere Island, most probably in the crystalline Pearya Terranes (Trettin, 1987). This allows us to draw the first ice drift vector for the Arctic Ocean: north of Ellesmere Island the drift of ice and icebergs was strictly to the east, towards Fram Strait.

In core Fl-561 the crystalline rocks alternate with quartz in abundance (Fig. 6.4). Whenever the quartz percentages are high, the crystalline rocks are low and vice versa. During the lower and older part of the core, quartz dominates more frequently than above, when crystalline rocks take over as the most important IRD constituent. Five of the coarse fraction peaks in the lower half of the core, peaks 7, 9, 10, 12 and 13 (Fig. 6.3) are also quartz peaks (Fig. 6.4). This indicates that during these older glaciations, the icebergs came from an ice complex that was grounded on a quartz-rich substratum. The nearest area with matching lithologic composition is the western part of the QEI with Mesozoic clastics and Cenozoic quartz sands. Icebergs from there could drift by northern Ellesmere Island following the same eastward drift track reconstructed in the previous paragraph.

If the source of the quartz-rich IRD had been the western QEI, then the accompanying rock fragments that were transported along with the quartz grains

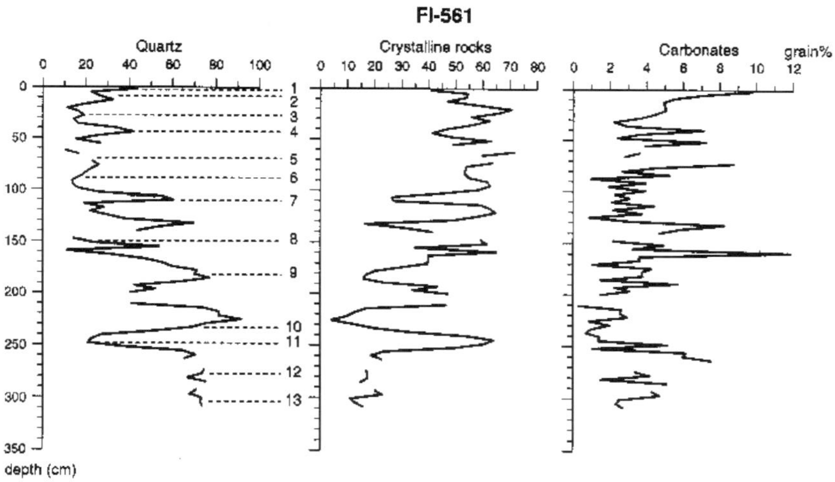

Fig. 6.4. Composition of IRD >250 μm in core Fl-561. Quartz shows overall a long term decline from values over 70% in the lower part of the core to around 30% near the top. Not only the average, but also the peak height decreases upsection. This trend is overprinted by shorter ups and downs with percentage differences of up to 60%. The quartz peaks in the lower half of the core are samples that, if looked at with the unassisted eye in their glass containers, give the appearance of nearly pure, clear quartz sand. In the long term, quartz is replaced by crystalline rocks from northern Ellesmere Island.

should be fine-grained clastic sedimentary rocks as they occur in the western QEI. Figure 6.5 shows that this was the case. In this figure, the percentages of clastic sedimentary rocks, crystalline rocks, and carbonates are converted into percent of all rock fragments, that is in percent of the non-quartz portion of the IRD. During the coarse fraction and quartz peaks 7, 10, 12 and 13, also the proportion of clastics among the rock fragments rose abruptly from very low values of less than 5% to as much as 35%. This is also true for the beginning and end of peak 9. The increases of the clastics clearly came at the expense of the crystalline rocks. Figure 6.5 also shows that every time that crystallines decrease, not only clastics, but also carbonates increase. There has been a shift from a more crystalline-dominated source area to another composed of sedimentary rocks and quartz.

The best explanation for these composition changes are shifts in the predominant iceberg sources from north to south within the high Canadian Arctic. During periods when crystalline rocks were high and quartz, clastics, and carbonates were low, icebergs came from outlet glaciers in the fjords of northern Ellesmere Island. During those times when the IRD consisted primarily of quartz, clastics and carbonates, the icebergs came from a glacier complex, or an ice sheet, in Axel Heiberg, Ellef Ringnes and adjacent islands in the northwestern QEI, the Innuitian Ice Sheet (IIS, Bischof and Darby, 1999).

This interpretation could be the answer to the controversy surrounding the IIS: it

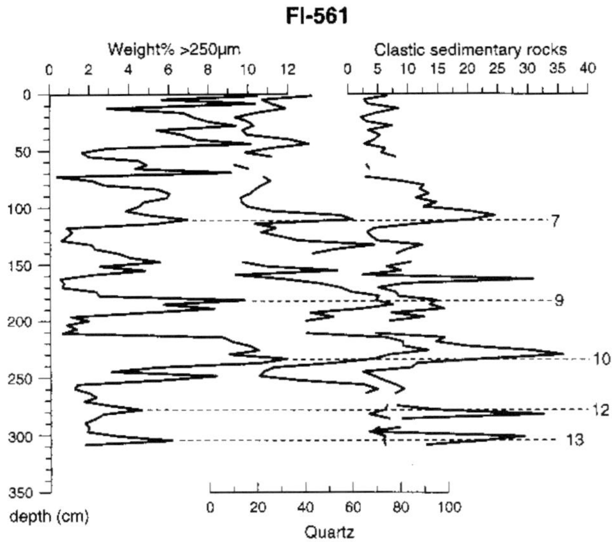

Fig. 6.5(a). Weight percent >250 μm IRD (left), its quartz content (middle), and the proportion of clastic sedimentary rocks among the rock fragments. Clearly, during the glaciations 7, 9, 10, 12 and 13, very coarse, quartz and clastic sedimentary rock rich IRD was carried by icebergs from the Innuitian Ice Sheet to at least north of Ellesmere Island.

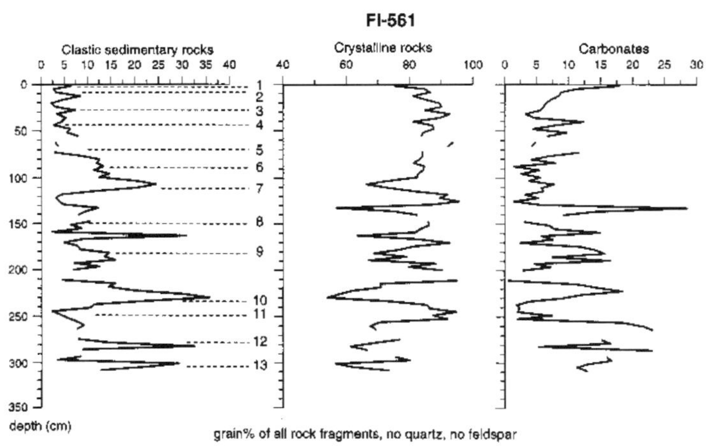

(b) IRD composition converted into percent of all rock fragments (total IRD minus quartz and feldspar mineral grains). The very coarse peaks 7, 10, 12 and 13, with some reservation also 9, are also quartz peaks and coincide with peaks of clastic sedimentary rocks. Therefore, most of the quartz was derived from a clastic sedimentary terrain. The proportion of clastic rocks among IRD rock fragments oscillates strongly between lows near 2% and maxima of 35%. Crystalline lithologies, however, are the most abundant IRD lithology and account for more than 50% in general, but during crystalline rock peaks, their proportion can exceed 90%. The average crystalline rock proportion is 70–80%.

has long been argued that the presence of crystalline shield erratics on the outer, Arctic-facing islands of the Canadian Arctic Archipelago required a massive glaciation over most of the archipelago (Tozer and Thorsteinsson, 1964; St. Onge, 1965), but geologic and geomorphic evidence that would support such a massive ice sheet such as glacial striations, till deposits, terminal moraines, etc. are conspicuously missing. Recently this controversy has been summarised by Arthur Dyke in a review article (Dyke, 1999), and the prevailing view now is that an IIS existed during the Late Wisconsinan Glaciation, but that it was not quite as extensive as previously thought. Dennis Darby and I have recently added to this debate by suggesting that a marine-based ice stream or glacier lobe must have extended from an ice divide to the northwest of Devon Island (Dyke, 1998) through Peary Channel to the Arctic coast (Bischof and Darby, 1999). The data of core Fl-561 demonstrate that previous editions of the IIS existed during older glaciations, in this case glaciations number 7, 9, 10, 12 and 13 (Fig. 6.5b).

6.2.2 The carbonate to quartz transition

The most important element of the IRD composition in the central Arctic Ocean are the changes from carbonate-dominated to quartz-dominated IRD. Depending on the sedimentation rates and the core penetration depth, the two types of IRD alternate repeatedly (Fig. 6.6). Of all cores analysed thus far, Fl-436 from the Alpha Ridge, central Arctic Ocean, shows the most pronounced alternations between carbonate and quartz dominated IRD. Both curves look like virtual mirror images of each other. Increasing proportions of quartz are balanced by decreasing proportions of carbonates and vice versa. The upper portion of the core is rich in carbonate and low in quartz. The other IRD lithologies are miscellaneous crystalline and clastic sedimentary rocks. Underneath follows a 30 cm thick unit in which the IRD consists almost exclusively of clear quartz. Below that level follow several more switches between carbonate and quartz-rich IRD. This alternation is the pre-eminent pattern of the IRD composition in the central Arctic Ocean and can also be seen in other cores from the same general area (Fig. 6.6b; Bischof et al., 1996; Bischof and Darby, 1997; Nørgaard-Pedersen et al., 1998). Therefore the sources of IRD switched between two major regions: one composed mostly of carbonate bedrock, the SW Canadian Arctic Archipelago, and another with predominantly quartz-rich substratum.

By virtue of new, ^{14}C AMS dated box cores from the central Arctic Ocean (Darby et al., 1997), the chronology of at least the last transition from quartz-dominated to carbonate-dominated IRD is known (Fig. 6.7). Before 45 ka, the IRD was dominated by quartz; the transition occurred between 45 and 40 ka, and after 40 ka, primarily carbonate-rich IRD from Arctic Canada was delivered to the central Arctic Ocean and beyond. Prior to the last glaciation, at least five older carbonate-to-quartz transitions (resp. carbonate peaks) occurred during the last 1 million years (Bischof et al., 1996).

The technique with which we (Dennis Darby and I) determine the origin of IRD is explained in detail in Darby and Bischof (1996).[25] It is a mathematical comparison of the IRD in any deep-sea sediment sample with the composition of debris in more

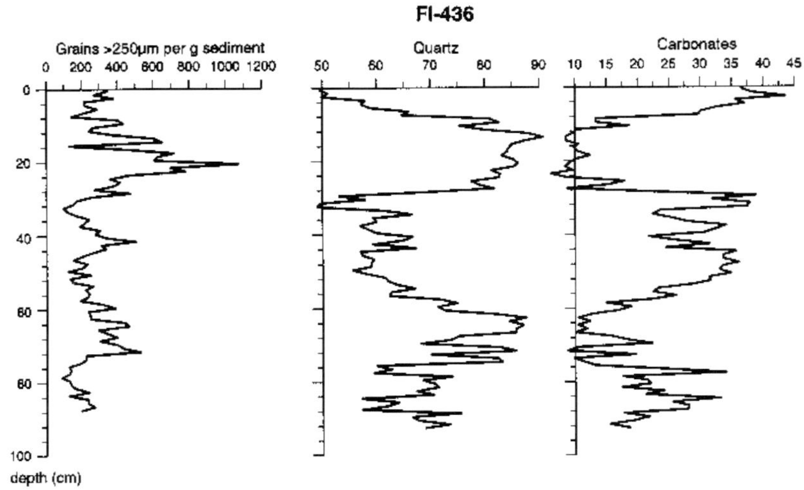

Fig. 6.6(a). Composition of IRD in core Fl-436, Alpha Ridge, central Arctic Ocean. Note the repeated alternations between carbonate- and quartz-dominted IRD. The quartz and carbonate curves look like mirror images of each other: whenever quartz is abundant, carbonates are low and vice versa. This demonstrates that this core site, and other places in the central Arctic Ocean, has received IRD from essentially two different sources: one that is carbonate-dominated (the southwestern Canadian Arctic Archipelago), and another one where quartz is abundant.

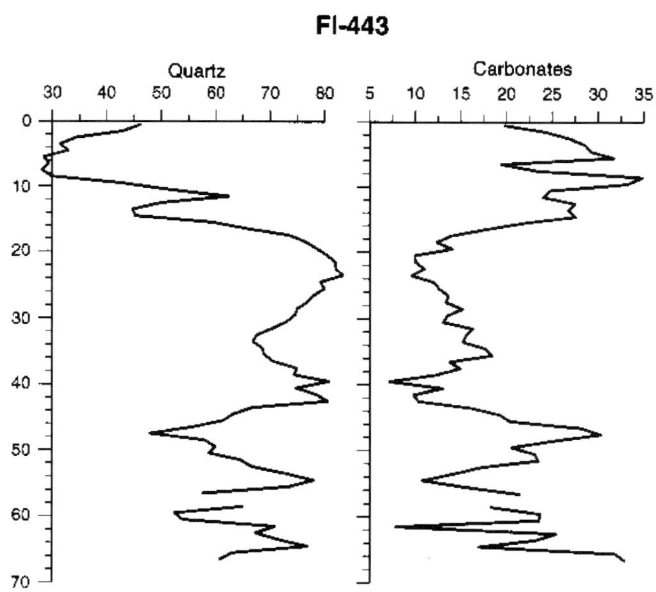

(b) IRD composition of core Fl-443, also Alpha Ridge. Apart from slight differences in the discrete percentages, this core almost exactly replicates the pattern of core Fl-436.

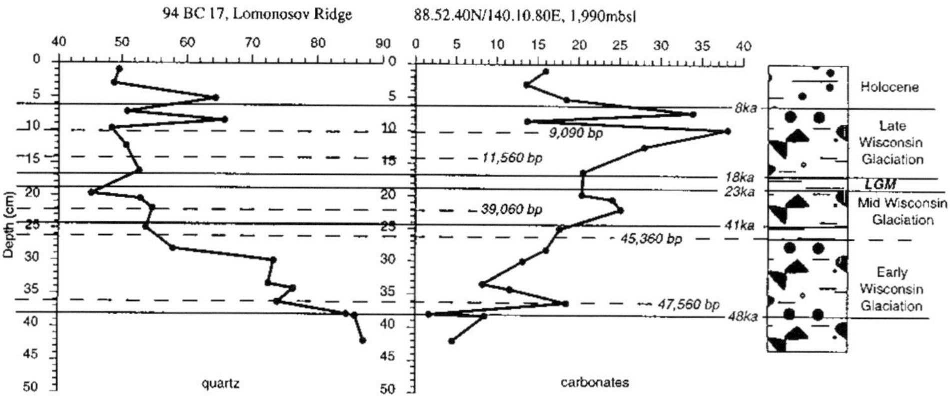

Fig. 6.7(a). Quartz to carbonate transition, looking upward, in IRD of the central Arctic Ocean.

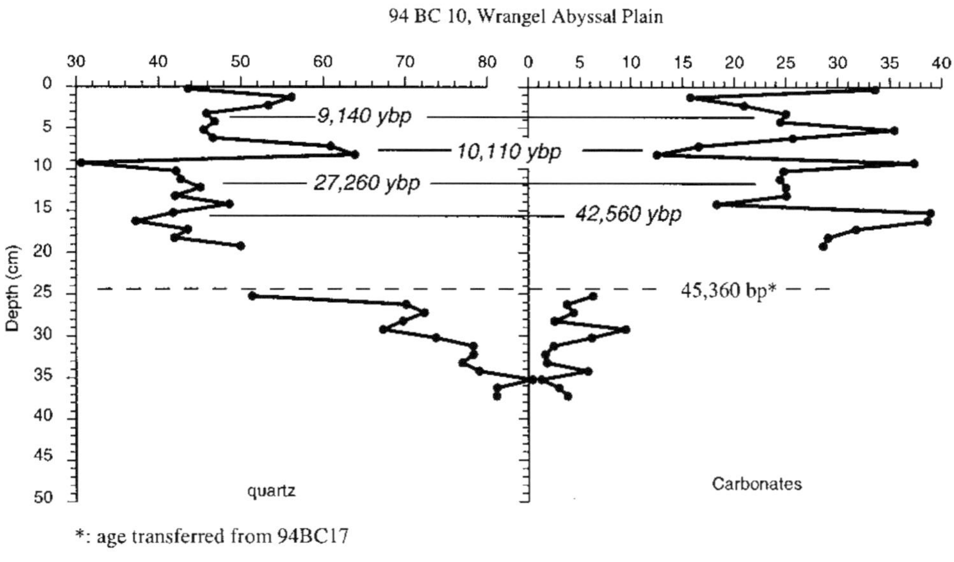

(b) 94BC17, fig. 41b: 94BC10. Quartz is initially dominant with percentages of over 90. These assemblages have a clear, bright appearance. Upsection, quartz is gradually replaced by carbonate IRD from the Canadian Arctic Archipelago. The chronology of this transition is constrained by several radiocarbon age determinations. However, one must be cautious with the ages because the cores have low sedimentation rates and have been affected by bioturbation. Also, the dates from the deepest parts of both cores approach the technical limits of the radiocarbon dating method. Despite these uncertainties, the lower, quartz-dominated IRD unit has a Middle-Wisconsinan age of over 40 ka, the quartz to carbonate transition occurred probably between 45 and 40 ka, and carbonate-rich IRD (with a minimum of 20 grain%) was delivered after 40 ka.

than 300 circum-Arctic source samples. The program output is a source group assignment and a probability. In general, when this probability is 0.95 or more, the sample's composition matches the data range of the assigned source very closely. In the case of some of the carbonate-rich IRD assemblages of the western and central Arctic Ocean, the values of all 80 variables were fully within the data range of the source group. Trustworthy source determinations were obtained for more than 70% of the samples. Carbonate and carbonate–clastic IRD assemblages were placed in the vicinity of Victoria Island in the Canadian Arctic Archipelago, and the sources for the quartz-rich IRD in the central Arctic Ocean are the Kara Sea and surrounding areas. The iron oxide grains in the 45–250 μm range, whose geochemical composition (12 elements) was used for source determination, always contain grains from multiple sources, often from both sides of the Arctic Ocean, in every sample. Usually, at least seven different source areas contributed to every sample. Thus, while one of the two major source areas, the Kara Sea and the southwestern Canadian Arctic islands, dominated alternately, several other minor source areas, including the Queen Elizabeth Islands, North Greenland, Svalbard, and the Laptev Sea also contributed to the IRD inventory. This indicates that the ice drift directions changed between two different modes during the time of deposition of every sample; at the low sedimentation rates in the central Arctic Ocean approximately every 1000 years. The Kara Sea dominance prior to 40 ka is in good agreement with the fact that the extent of the Kara Ice Sheet (KIS) was largest during the early and middle Weichselian (Thiede and Mangerud, 1999). Thereafter the late Wisconsinan LIS took over as the major IRD source. Figure 6.8 (see colour section) summarises the identified source regions of IRD and the reconstructed driftpaths from the middle to late Weichselian (Wisconsinan) Glaciation to the present.[26] On the North American side, the Laurentide and Innuitian ice sheets, and on the Eurasian side the Barents/Svalbard and Kara ice sheets delivered most of the IRD.

6.2.3 Direction of ice drift and surface circulation

The circulation necessary to disperse carbonate-rich IRD from the southwestern Canadian Arctic islands in the western and central Arctic Ocean is shown in Fig. 6.9a. I like to call this circulation the carbonate mode, because in this mode, large quantities of icebergs charged with carbonate clasts (and clastic sedimentary and crystalline rocks) drifted from the northwestern LIS in a wide arch across the central Arctic Ocean, finally leaving through Fram Strait. In this mode, the eastern Arctic Ocean receives debris exclusively from northern Eurasia. A very large part of the Arctic Ocean exports ice into the Norwegian Greenland Sea, and the flow direction is southward across the entire width of Fram Strait. There was no northward surface flow into the Arctic Ocean. When this flow regime was active during a glacial stage, pack ice and icebergs with sedimentary rocks and coal from the East Siberian Sea could pass through eastern Fram Strait and continue drifting southward through the central NGS, as it has been demonstrated for stage 6 sediments (see Section 5.10, and Bischof et al., 1990). Similarly, during interglacial times when the large ice sheets were absent, or at least much smaller than during the glacials, sea ice and icebergs from around Svalbard and sea ice from coastal areas of the East Siberian Sea with

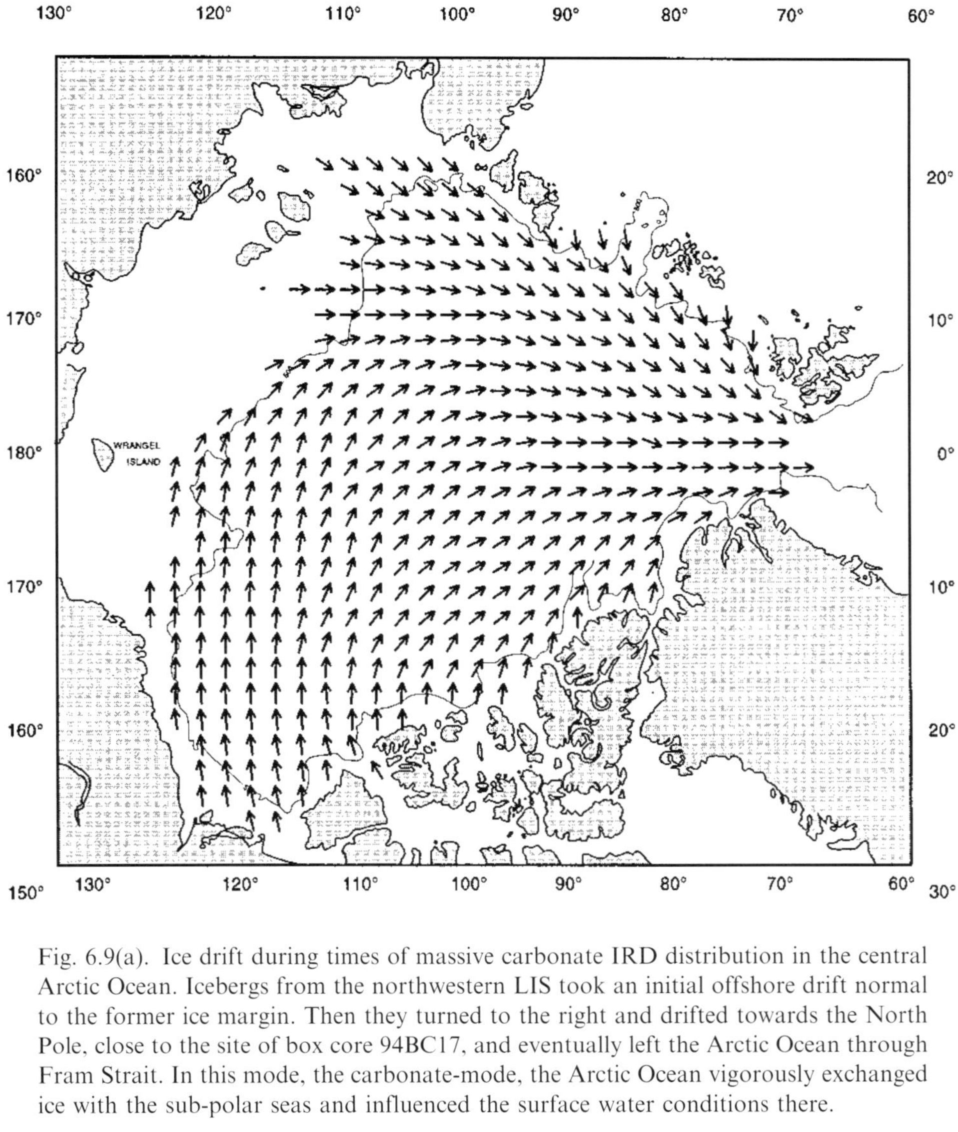

Fig. 6.9(a). Ice drift during times of massive carbonate IRD distribution in the central Arctic Ocean. Icebergs from the northwestern LIS took an initial offshore drift normal to the former ice margin. Then they turned to the right and drifted towards the North Pole, close to the site of box core 94BC17, and eventually left the Arctic Ocean through Fram Strait. In this mode, the carbonate-mode, the Arctic Ocean vigorously exchanged ice with the sub-polar seas and influenced the surface water conditions there.

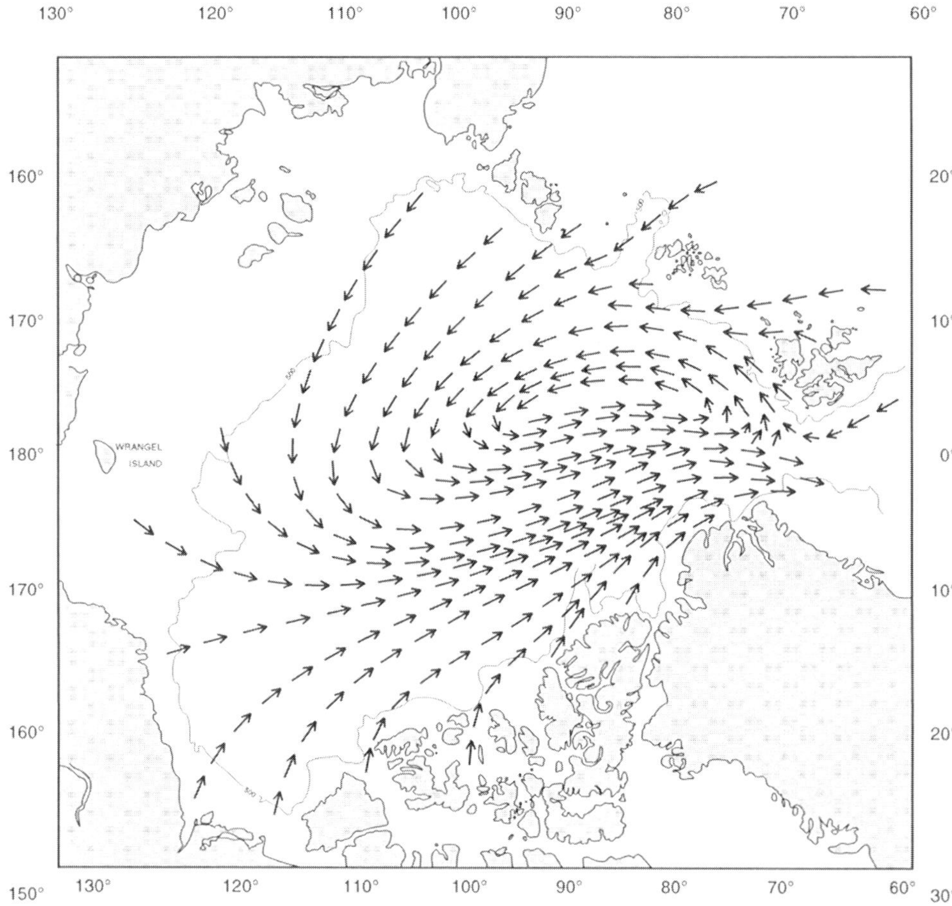

(b) The quartz mode of the Arctic circulation. In this mode, quartz-rich IRD from the Kara Ice Sheet is dispersed in a counterclockwise gyral motion in the central Arctic Ocean. The western Arctic Ocean and the part close to North America were not affected by the quartz-rich IRD. In this mode, ice could remain trapped in the eastern Arctic Ocean. There also was the possibility of a surface flow into the Arctic Ocean through eastern Fram Strait, as documented by Spielhagen with chalk (Spielhagen, 1990).

coal could drift into the eastern Norwegian Sea and significantly expand the sea ice cover (see Section 5.4.4).

Figure 6.9b illustrates the other state of the Arctic circulation, the quartz-mode. In this mode, IRD very rich in quartz from the Kara Ice Sheet is dispersed in the central Arctic Ocean in a counterclockwise gyre. This is a less communicative state in which the Arctic Ocean retains some of its ice for possibly extended periods of time. The main outflow is through western Fram Strait, which becomes the thoroughfare of a mixture of icebergs from both the Laurentide and Innuitian ice sheets. There was a surface flow into the Arctic Ocean through the eastern Fram Strait, as proposed for oxygen isotope stage 2 by Spielhagen (1990).

Both modes of the Arctic circulation alternated on time scales shorter than the time of deposition of the analysed samples, within less than 1000 years. This interpretation is based on the occurrence of iron oxide minerals from multiple sources, sometimes on both sides of the Arctic Ocean, in the same samples. However, some cores from the western Arctic Ocean have higher and variable sedimentation rates of up to 6 cm/kyr (Darby et al., 1997). For these intervals, every 1 cm thick sample represents approximately 170 years. Therefore, the two circulation systems may have alternated on time scales as short as multi-decadal.

The long-term trend from a middle Weichselian dominance of the quartz-mode (from the Kara Ice Sheet) towards a late Wisconsinan dominance of the carbonate-mode (from the Laurentide and Innuitian Ice Sheets) occurred in response to sequential ice build-up. During the early Weichselian, ice and glaciers began to grow in and around Svalbard, as indicated by large amounts of clastic sedimentary rock IRD from the Barents Sea during the early Weichselian in kastenlot core 23071 from the Norwegian Sea (see Section 5.8.1). Sometime during the middle Weichselian the ice build-up has produced a large, combined Barents–Kara Ice Sheet (Thiede and Mangerud, 1999). The ice mass must have had a sufficiently strong and stable high atmospheric pressure cell to drive the circulation of the quartz-mode. Ironically, the inflow of relatively warm surface waters around Svalbard into the Arctic Ocean during the early Weichselian may have caused the glacier growth via moisture supply. As the Barents and Kara ice sheets grew, increasing the albedo, the average temperature over the high Arctic dropped. At one point, the summer melting period was no longer strong enough to remove the snow from the previous winter. Consequently, glacier growth began in northern North America, finally leading to the large Laurentide and smaller Innuitian Ice Sheets. As the ice mass of the Kara Ice Sheet shrank because of iceberg calving and meltwater discharges, the North American ice sheets grew to maximum extent, and the main control force of the surface circulation shifted to North America. The carbonate-mode took over, and the drift of the icebergs was controlled by high atmospheric pressure over the Laurentide and Innuitian ice sheets.

Another climatically intriguing aspect of the quartz-mode (Fig. 6.9b) is that, while being in this mode, the Canadian Arctic Archipelago could receive a lot of moisture to feed glacier growth from the North Pacific. The central part of the Gulf of Alaska was devoid of IRD (St. John and Krissek, 1999) and thus was open water even during the last glaciation. At the present, the Arctic islands receive very little moisture (15 cm/year), and it would take thousands of years of accumulation to build a sizeable ice sheet. But with an increasing supply of moisture from an open water North Pacific Ocean, the North American ice sheets could have grown much faster. The growth of the Kara Ice Sheet, and the change of the atmospheric circulation that it caused (opening a gateway for moisture from the North Pacific), could have been responsible for the growth of the Laurentide and Innuitian ice sheets.

6.2.4 Decay of the Laurentide and Innuitian Ice Sheets

How did the disintegration of the two large North American ice sheets, the Laurentide Ice Sheet (LIS) and the Innuitian Ice Sheet (IIS) proceed? How many icebergs were in the Arctic Ocean at different times, and how much freshwater equivalent volume was exported into the sub-Arctic seas?

The surface area covered by both LIS and IIS was approximately 1.3 x 10^6 km^2. At an average ice thickness of 600 m, the total ice volume at the time of maximum ice advance was 7.7 x 10^5 km^3. This volume, plus the amount that accumulated between the LGM (18 ka) and the time when ice retreated onto land (9 ka), was available for iceberg and meltwater production within a 9000 year deglacial period. At an average accumulation rate of 15 cm water per year, which is the present-day average over the high Canadian Arctic, and under the assumption of complete accumulation (= ice formation), approximately 1.9 x 10^6 km^3 would have accumulated in addition to the ice that already existed at the LGM. The total volume of ice that was lost from the LIS and IIS therefore was 2.7 x 10^6 km^3. This translates into 300 km^3 of ice that were discharged annually into the Arctic Ocean. In order to calculate how many icebergs resulted from disintegration, we must make an assumption about an iceberg's size. Modern day analogies from Antarctica, West Greenland and the Scoresby Sund fjord system of East Greenland suggest that a size of 1 km diameter and an average thickness of 400 m are quite realistic for an 'average' Late Pleistocene Arctic Ocean iceberg. At these measures, our idealised iceberg had a volume of 3.14 x 10^8 m^3, and the annually discharged amount of ice would produce 960 icebergs of the above dimension. But this is a maximum figure that assumes no meltwater production. If we factor in that approximately 20% of the deglaciation was in the form of meltwater, then the iceberg frequency drops to 770 icebergs per year.

At an average drift velocity of 5 cm/sec, and following the drift track of the carbonate mode (Fig. 6.9a), the icebergs could drift as far as 1600 km from the ice margin in one year. But because the ice drift is not straight and linear, but rather crooked and irregular, the true distance that the icebergs could have travelled is less, perhaps 1200 km. (During the SHEBA experiment, see endnote 3, the research vessel Des Groseillers drifted 800 km away from the position where it was allowed to be frozen into the Arctic ice pack; during the same time of a little over one year, the total drift distance was 2700 km, (Perovich et al., 1999)). Therefore, after one year of ice drift, a swath of the western Arctic Ocean 1200 km long and approximately 600 km wide would be occupied by approximately 800 icebergs. This is equivalent to one iceberg in 900 km^2, or one iceberg every 30 km. But since many icebergs are smaller than our standard size iceberg, and because many of the larger icebergs break up into smaller pieces, the total number of icebergs in the Arctic Ocean was most probably larger than 800 per year. By analogy to iceberg observations by Dowdeswell et al. (1992), the number may be closer to 8000 per year. In that case there would have been one iceberg every 9.5 km. This density is similar to that of icebergs in the winter in the Labrador Sea (personal observation during a flight across southern Greenland and Baffin Island). Therefore, the average Late Pleistocene 'iceberg traffic' in the Arctic Ocean was similar to the present-day winter scenario in the Labrador Sea.

But these values were calculated for an average scenario. In reality, the iceberg frequencies and the velocity of Laurentide and Innuitian ice sheet retreat varied through time. If we regard the densities of IRD grains in the sediment as an approximate measure of the frequency of icebergs that traversed each core site, then the average frequency of one iceberg every 9.5 km equals the average Late Pleistocene grain density of 152.8 IRD grains >250 μm per gram sediment.[26] Figure 6.10 shows the grain density fluctuations in box core 92BC6. Aside from an extreme maximum of 1353 grains/g sed., the values in the Pleistocene section (below 10 cm) are usually in the range of 100–300 grains/g sed. The equivalent iceberg density for a value of 100 grains/g sed. is one iceberg every 11.7 km, and for 300 grains/g sed. it is one every 6.8 km. Therefore, icebergs in the Arctic Ocean during the Late Pleistocene deglaciation were spaced between 6 and 12 km apart. These distances are similar to those close to the large Antarctic ice shelves (personal observation made by Bettina Fach, Center for Coastal Physical Oceanography, Norfolk, Virginia). Therefore we can envision the Late Pleistocene Arctic Ocean between 18 and 9 ka as permanently ice covered and dotted with icebergs 6 to 12 km apart.

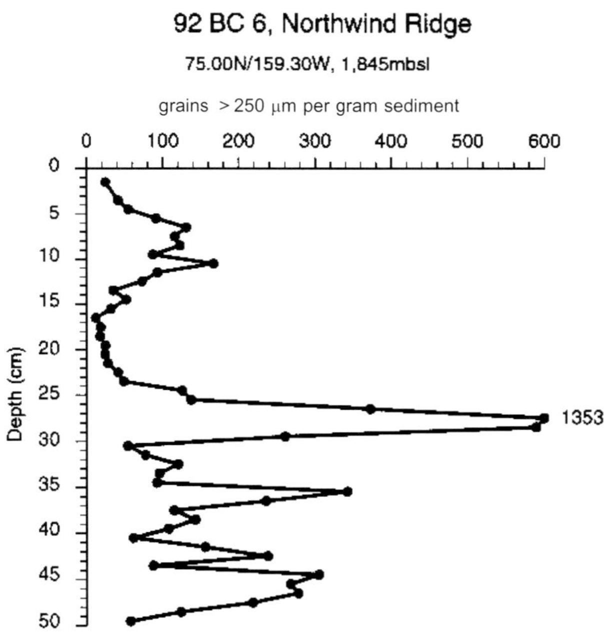

Fig. 6.10. IRD grain densities in box core 92BC6. The maximum density is 1353 grains >250 μm per gram sediment, but during the Pleistocene section below 10 cm core depth, the values are generally in the range of 100–300 grains per gram sediment.

6.3 FRESHWATER SUPPLY

The amount of freshwater that was delivered by icebergs from the Laurentide and Innuitian Ice Sheets to the Greenland Iceland and Norwegian (GIN) seas and its importance for the watermass stratification will be discussed next. In contrast to the present Holocene warm phase when meltwater is brought to the Arctic Ocean by the large rivers in Siberia and the Mackenzie River, the meltwater delivery during the Pleistocene was profoundly different. Large amounts of freshwater were stored in the icebergs and the Pleistocene pack ice and drifted into the GIN seas. There, the ice melted and lowered the surface water salinity.

The total volume of ice from glacial icebergs from the LIS and the IIS between 18 and 9 ka was 2.7 million cubic kilometres. This translates into 2.43 x 10^6 km^3 of water if a density of 0.9 g/cm^3 for ice is used. The discharge was 270 km^3 per year at a range from 180–530 km^3 per year. In addition to this, the sea ice within the sector of the Arctic Ocean in which the icebergs drifted also moved towards the GIN seas. Because of the colder temperatures during the Pleistocene, the sea ice was very probably thicker than today. At an average thickness of 5 m and the same area (1200 x 600 km), the freshwater equivalent from sea ice rafting was 3600 km^3 per year. Therefore the freshwater contribution from sea ice was more than 10 times larger than the one resulting from melting icebergs. This brings the total freshwater volume from icebergs and sea ice to 3870 km^3 per year, with a minimum of 3780 and a maximum of 4130 km^3 per year.

The freshwater flux to the GIN seas was 122,700 m^3 per second with a range from 120,000–131,000 m^3/sec. To put these values in perspective, the annual Pleistocene freshwater flux from icebergs and sea ice from the western Arctic Ocean amounts to 0.12 Sverdrup (1 Sv = 1 million cubic metres of water per second), very small compared to the recent watermass exchange through Fram Strait of circa 3 Sv going in and 3 Sv flowing out of the Arctic Ocean. But how does the Pleistocene freshwater transport of icebergs and sea ice compare to the amount of freshwater in modern sea ice as it drifts from the Arctic Ocean into the Greenland Sea? The amount of freshwater contained in sea ice in the same sector occupied by the Pleistocene icebergs at an average thickness of 3 m is 2160 km^3 per year. This theoretical value compares favouribly with modern satellite-based calculations of the annual sea ice volume and aerial flux. Kwok and Rothrock (1999) calculated the annual (12 month) volume flux to be 2366 km^3 (my value: 2160 km^3), and the area flux to be 919,000 km^2 (my value: 720,000 km^2). Thus, the amount of freshwater that was delivered from the Arctic Ocean to the sub-Arctic seas during the Late Pleistocene by sea ice and glacial icebergs was 80% larger than at present (3870 km^3 per year, minimum: 3780, maximum: 4130 km^3 per year).

It should be borne in mind that these values are minimum figures, calculated for the contributions from the North American ice sheets only. There were other sources of glacial ice in and around Svalbard and North Greenland that contributed to the Pleistocene freshwater budget of the GIN seas (Fig. 6.8). These additional sources of glacial ice would increase the amounts of freshwater from icebergs, perhaps to values in the order of 300–400 km^3 per year. For this reason, the total amount of freshwater that moved southward in form of ice through Fram Strait during the Late Pleistocene was almost certainly more than twice the amount carried by modern sea ice.

6.3.1 Surface circulation

The movement of icebergs and pack ice from the LIS and IIS towards Fram Strait must have exerted an influence on the surface waters which were in direct contact with the ice. Most likely the surface waters down to a depth of the largest iceberg keels (400 m) moved in approximately the same direction as the ice. At least this is what drifting buoys with current meters anchored in pack ice confirm for the upper 250 m of the water column (figure 5 on page 114 in Johannessen et al., 1994). If we assume that the upper 400 m of the Arctic Ocean waters followed the same path as the drifting icebergs, then 9.13 x 10^6 m^3/sec flowed out of the Arctic Ocean into the Greenland Sea. These are 9 Sv as opposed to only 3 Sv today. This implies that the Late Pleistocene surface current from the Arctic Ocean was three times stronger than the present-day East Greenland Current.

If the surface circulation had been sluggish and slow, the icebergs could stay in the Arctic Ocean for extended periods of time, possibly several decades as ice island T-3 did (Clark et al., 1980). Undergoing multiple summer melting seasons, the icebergs from the LIS and IIS had plenty of opportunities to melt and release their ice rafted debris to the sea floor. The sedimentation rates therefore should be relatively high. But the opposite is true. Directly in the area affected by the carbonate-mode circulation, sedimentation rates are the lowest of any core analysed thus far (Bischof and Darby, 1997), only fractions of a millimetre per thousand years. These low rates are better explained by a stronger and faster flow towards Fram Strait rather than by sluggish circulation.

6.3.2 Influence on the biology

But the carbonate-mode of the Arctic circulation with a strong surface outflow and a compensatory inflow at greater depth could explain in a simple yet elegant way another characteristic of Arctic Ocean sediments. Sediments older than 15.7 ka (calendar) in the western Arctic Ocean are devoid of marine microfauna remains (Darby et al., 1997). In the eastern Arctic Ocean during the same time period, the abundances of microfossils (foraminifera) drop from relatively high abundances like those of the Holocene before approximately 38 ka (exact timing is uncertain) to very low values before rising again at 15.7 ka (Darby et al., 1997). As the foraminifera abundances increase after 15.7 ka, the amounts of coarse IRD drop to low Holocene values (Fig. 6.11). Most micropaleontologists explain the foraminifera-barren unit as a section that was affected by carbonate dissolution, which results from oxidation of organic matter near the bottom of the ocean. This process produces the carbonate-aggressive anion HCO_3^- which then dissolves the carbonate shells of the microorganisms.

I find this proposition for the Arctic Ocean unconvincing and suggest a different, physical process, that effectively removed all surface water biota from the western Arctic Ocean, making it a Pleistocene biological desert. If carbonate dissolution really had been operative, it should have affected the detrital, ice rafted carbonate clasts as well, but these are particularly abundant in the strata where the foraminifera are missing. And this is not only true for the >250 μm fraction but also for the finer sand and silt fractions. All these fractions contain abundant carbonate. Also, the detrital carbonate clasts do not show any signs of dissolution such as cavities.

The only way to save the dissolution theory would be to argue that the bottom

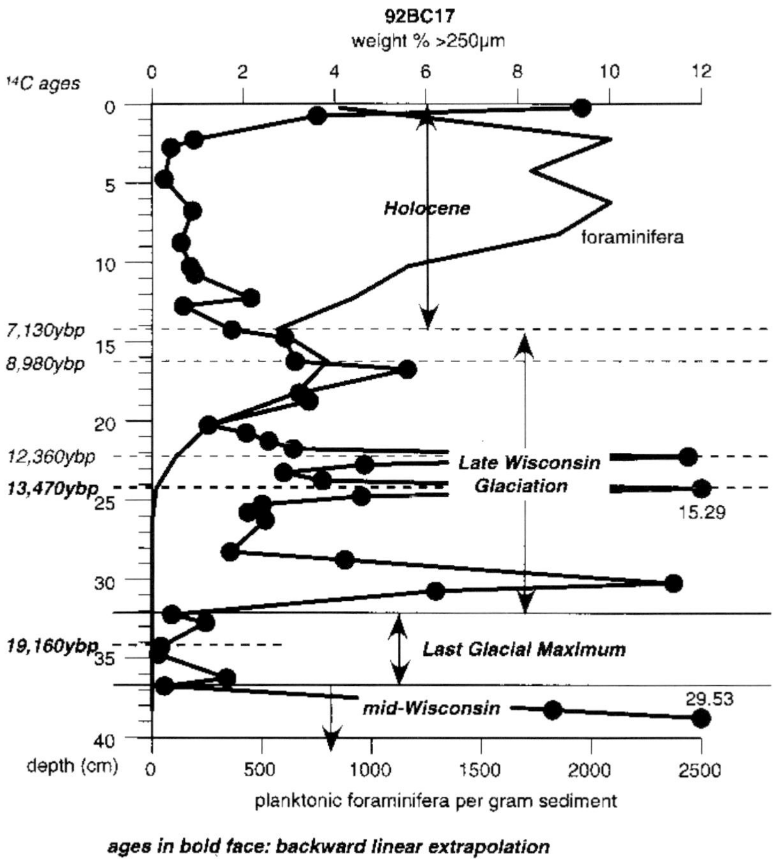

Fig. 6.11. Age model of the Late Wisconsinan to Holocene on the basis of core 92BC17. Ages to the left are actual ^{14}C AMS age measurements. Two ages in bold print are linear extrapolated. None of the ages are converted into calendar years. A key stratigraphic horizon in all cores from the western and central Arctic Ocean is the 13.5 ka level where foraminifera reappear after a prolonged absence and where the amounts of Pleistocene ice rafted debris drop.

waters were barely acidic enough to dissolve foraminifera shells but not detrital carbonate clasts. But in that case the bottom waters should quickly become carbonate-saturated, and therefore any following foraminifera tests should still accumulate.

My proposition is that the carbonate-mode was a wide surface current that carried icebergs from primarily the LIS, surface waters, and all marine microorganisms that lived floating in it, out of the Arctic Ocean. The surface current was balanced by an inflow at greater depth through Fram Strait which carried either few microorganisms or none at all. This undercurrent occasionally turned into a surface current when the density difference to the ambient surface waters disappeared (according to a process described in Section 5.9). This surface current flowed northward through either the eastern Fram Strait or across the Barents Sea when it was not fully blocked by glacial ice, and carried microorganisms from the NGS into the eastern Arctic Ocean, which explains the continued presence of microorganisms there at low abundances while they were absent in the western part. At the same time, most of the marine connections between the Arctic Ocean and the sub-Arctic seas from where biota could have been imported (Bering Strait, channels in the Canadian Arctic Archipelago, Nares Strait and the Barents-Kara seas) were closed owing to the lower sea level or blocked by the Pleistocene glaciers.

Exactly at the level where the foraminifera reappear or sharply increase in abundance, the IRD composition switches briefly from carbonate-dominated to quartz dominated (Fig. 6.12). The sources of the quartz-rich IRD are in the Kara Sea, and the very brief peak marks a widening of the quartz-mode into the western Arctic Ocean. The reappearance of the foraminifera is thus closely linked to this westward expansion of the quartz-mode. This change of the surface ocean circulation occurred at 15.7 ka. Therefore it is slightly younger than the northward protrusion of temperate surface waters into the Norwegian Sea, which began at 17.5 ka in connection with the Barents Ice Sheet decay. Approximately 1800 years later these northward moving waters reached into the western Arctic Ocean.

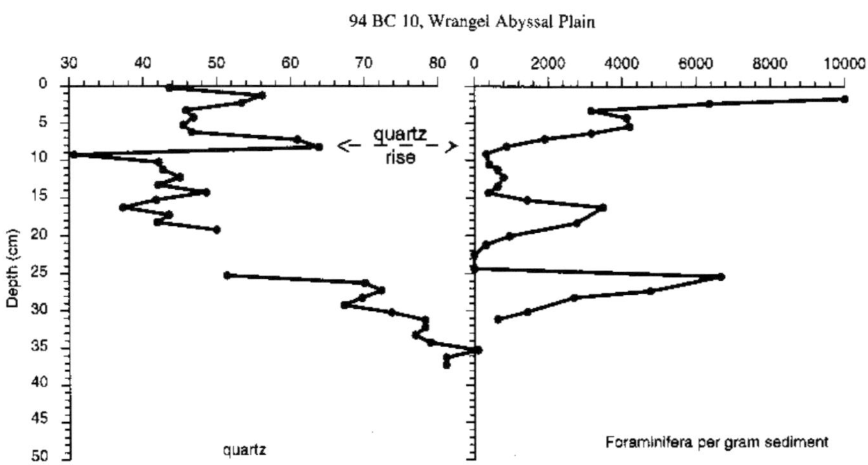

Fig. 6.12. Note the drastic rise of the quartz percentage in 94 BC 10 from its lowest values of 30% to more than twice that at the exact depth where the foraminifera abundances increase. The quartz-rich debris is from the Kara Sea and reflects a brief return to the quartz-mode of the Arctic circulation after a long dominance of the carbonate-mode. This change of the surface current opened the way for foraminifera to repopulate this part of the Arctic Ocean.

7

Sea ice motion: the physical foundations and implications

7.1 SEA ICE IN THE CLIMATE SYSTEM

Sea ice represents a thin, broken layer on the polar oceans. It acts as an effective boundary between the two much larger geophysical fluids, the atmosphere and the ocean, influencing their interaction considerably. Due to ocean tides and surface waves the sea ice is broken into floes of varying size ranging from metres to tens of kilometres. The evolution of the sea ice distribution is controlled by dynamic and thermodynamic processes which are driven mainly by the atmosphere and, to a lesser degree, by the ocean. Freezing and melting of sea ice are determined by the energy balance at the ocean surface, whereas the motion of the ice floes is mostly caused by the wind. Scattered within the pack ice are icebergs (broken off from glaciers) and ice-islands (broken off fast ice). In the Arctic their size is usually small with little draught, and they generally move with the sea ice pack. With respect to their motion icebergs and ice-islands can, therefore, be treated as sea ice. In the open ocean icebergs move with the oceanic surface currents which are dominated by the wind systems.

In the climate system sea ice plays an important role. Due to its high albedo it modifies the surface radiation balance, and it influences the exchange of momentum, heat and matter between atmosphere and ocean as a result of its insulating behaviour. Since the salinity of sea ice is on average only 5psu as compared to 35psu of sea water, the freezing of sea ice during cooling periods initiates brine expulsion and subsequent convection with deepening of the surface mixed layer and thus affects the formation of deep and bottom water. During the melting period relatively fresh water stratifies the oceanic surface layers, i.e. the oceanic mixed layer retreats to shallower depths.

As a result of the sea ice motion, the melting generally occurs in places far from the formation area. Averaged over a seasonal cycle, the net freezing rate in a certain area is therefore rarely zero, but rather positive or negative depending on the mean divergence or convergence of the sea ice flow. The net freezing rates represent strong surface buoyancy fluxes which heavily affect the density structure and therefore the thermohaline circulation of the ocean. The spatial and temporal characteristics of the sea ice motion depend on the rheological aspects of sea ice as a deformable plastic material, on the geometry of the coastlines, and on the atmospheric forcing fields, in particular the wind.

Sea ice covers a substantial part of the polar oceans. In the Arctic the average winter sea ice extent amounts to approximately 13 million km^2. The summer extent is about 7 million km^2. Superimposed on the seasonal cycle are pronounced anomaly fluctuations which on average are up to one-third of the seasonal amplitude. Similar to other geophysical variables the anomaly fluctuations can occur on a broad range of time scales with increasing amplitudes for longer periods, showing red variance spectra. The time series for sea ice extent which have been collected over both polar regions since the end of 1972 with satellite remote sensing techniques do not reveal any strong trend but show pronounced long-term fluctuations. Ice motion data on a larger scale, available since 1979 from the International Arctic Buoy Programme, also show a pronounced variability.

Unfortunately, there is no such information about the observed variability and trend for sea ice thickness. Since 1990 the sea ice thickness has been monitored with upward looking sonars in Fram Strait and a few other locations in the Arctic and Antarctic. These time series, however, are too short, and the spatial coverage is too small to significantly indicate long-term changes of the sea ice mass. Preliminary estimates of the observed sea ice export through Fram Strait from these data yield a mean value of about $0.1 Sv$ ($1 Sv \equiv 10^6$ m^3 s^{-1}) with strong fluctuations about this mean, as well as an increase from 1990 to 1994 (Vinje et al., 1997). The exported sea ice melts in the Greenland Sea and also south of Denmark Strait, thereby modifying the oceanic surface buoyancy flux which has strong implications for the formation of deep water and the oceanic thermohaline circulation in the North Atlantic. In some regions this freshwater flux from sea ice melt is more than twice the annual rain-rate.

7.2 OBSERVATIONS OF SEA ICE DRIFT

The first accounts of drifting sea ice originate from the early explorers of the northern polar seas. A detailed view was obtained from Nansen during his drift with the research vessel ‘Fram’ in 1893–1896 and from many drifting ice stations (see Fig. 7.1). The first drifting station was set up by the Soviets near the North Pole in May 1937 headed by I.D. Papanin. Moving rapidly south through Fram Strait into the Greenland Sea it was occupied until February 1938. Between 1950 and 1991, 30 more stations were set up by the Soviets.

Since 1979 the spatial pattern of the sea ice motion has been determined from drifting buoys which are deployed by the International Arctic Buoy Programme. The most advanced technique developed recently to obtain large-scale information on sea ice drift is the investigation of the displacement of brightness temperature patterns observed from consecutive images of passive microwave satellite sensors (Agnew and

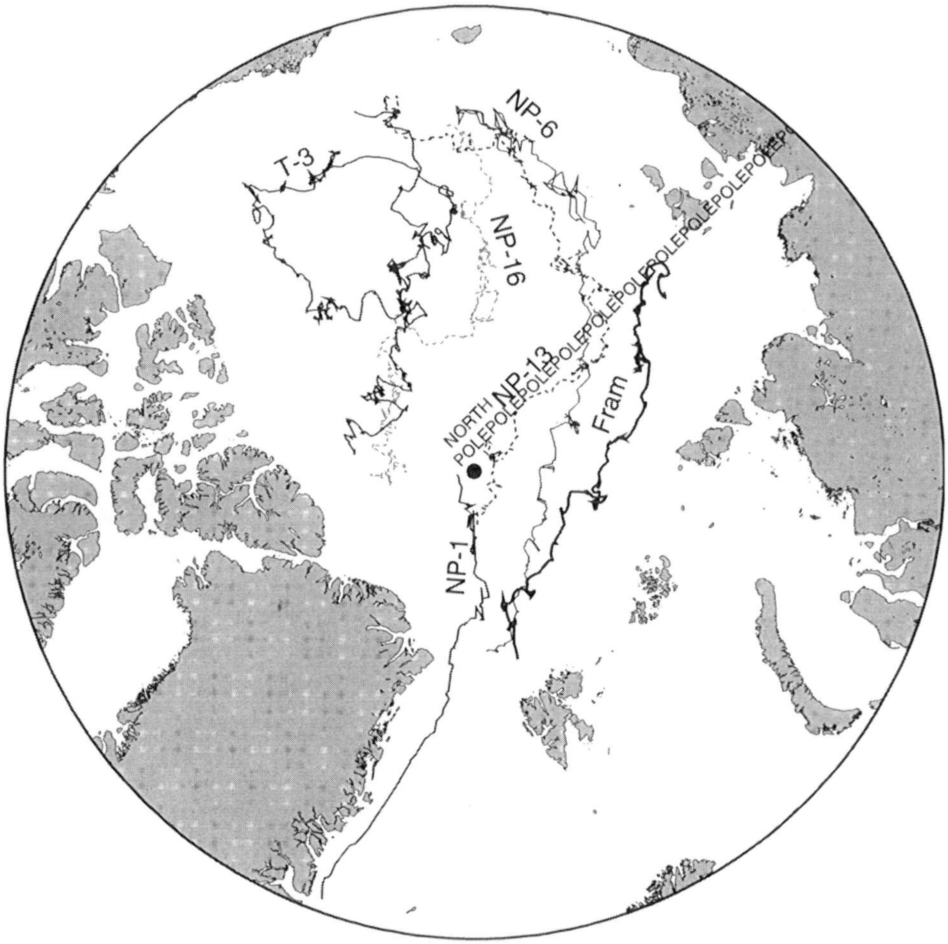

Fig. 7.1. Drift tracks of FRAM (1893–1896), a collection of Soviet North Pole Stations (NP–1: Papanin, 1937–1938; NP–6: 1956–1959; NP–13: 1964–1967; NP–16: 1968–1972) and the US Ice Island T–3 (1961–1971).

Le, 1996; Martin and Augstein, 2000). This technique has the advantage that global coverage can be obtained each day since the images are not obscured by clouds which are transparent to microwave radiation, and the horizontal resolution is of the order of 30 to 50 km. An example is shown in Fig. 7.2 which displays the sea ice motion between 13 and 16 January 1994.

The ice drift responds on time-scales of a few days to changes of the wind systems. Fig. 7.3 shows the ice motion between 17 and 20 January 1994 with a strongly changed drift pattern as compared to the situation three days before (Fig. 7.2).

Both Figs 7.2 and 7.3 indicate that drifting buoys and remote sensing data at the buoy position give similar information, i.e. the satellite algorithm is a precise and

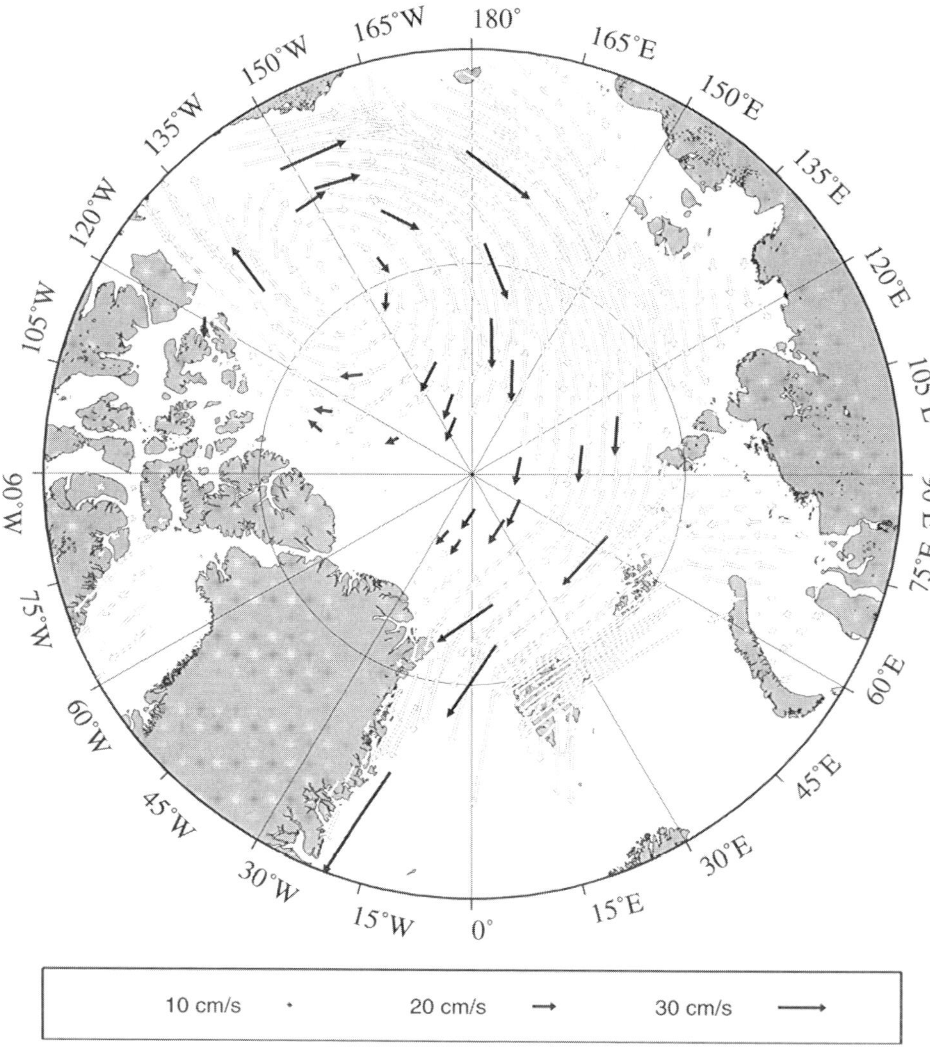

Fig. 7.2. Sea ice motion from 13–16 January 1994 determined from SSM/I images (light arrows) and from drifting buoys (dark arrows) (courtesy T. Martin, IfM, Kiel).

useful tool. Unfortunately, this technique is not applicable during the melting period, because the microwave emission from a wet surface can no longer be interpreted in terms of types or ice concentration.

The mean sea ice drift for January 1994 is displayed in Fig. 7.4. With a pronounced Beaufort Gyre and the Transpolar Driftstream this drift pattern is very similar to the long-term mean.

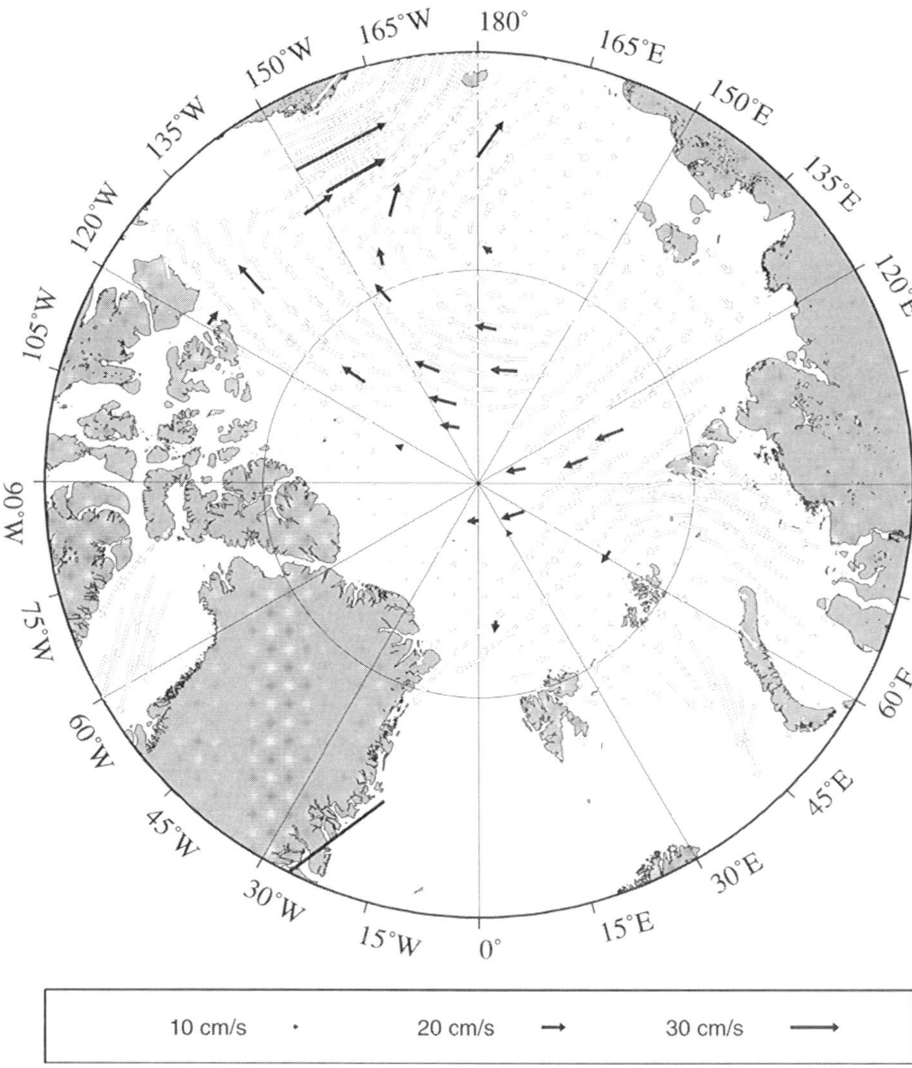

Fig. 7.3. Sea ice motion from 17–20 January 1994 determined from SSM/I images (light arrows) and from drifting buoys (dark arrows) (courtesy T. Martin, IfM, Kiel).

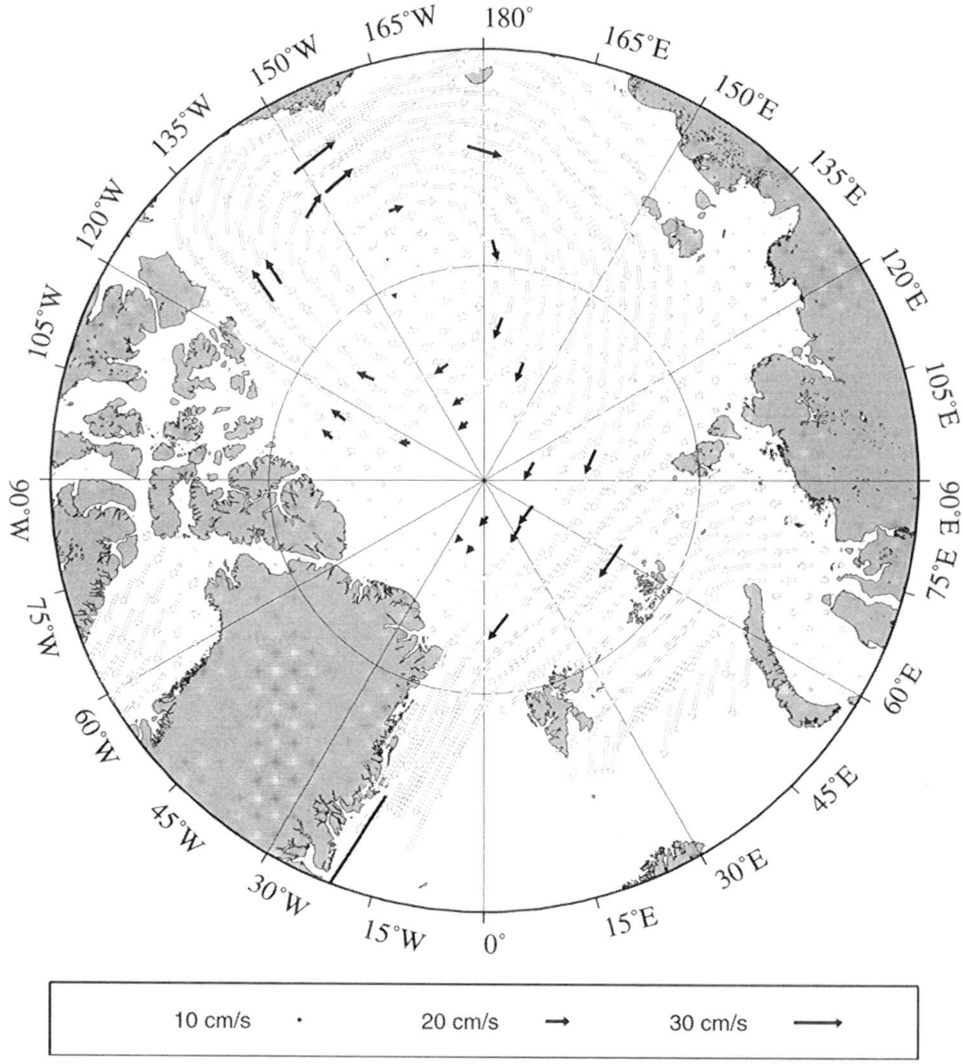

Fig. 7.4. Mean sea ice motion for January 1994 determined from SSM/I (light arrows) and from drifting buoys (dark arrows) (courtesy T. Martin, IfM, Kiel).

7.3 MODELS OF SEA ICE MOTION

For a long time, climate models have treated the sea ice as a non-moving slab with a thermodynamic response to atmospheric and oceanic energy fluxes. In recent years it has been found that the motion of the ice floes significantly modifies the surface forcing of the ocean. Because of the solid state properties of sea ice, the determination of the sea ice velocities is more difficult than for winds and ocean currents. In dynamic sea ice models it is generally assumed that the sea ice can be treated as a two-dimensional deformable continuum, which is characterised by five variables, i.e. the fields of ice and snow thickness, compactness (concentration), and horizontal velocity. The prognostic equations are derived from conservation equations for ice and snow mass and compactness, and from a momentum balance which includes internal ice stresses. Dynamic–thermodynamic sea ice models consist of four major components:

1. A surface energy balance which uses the atmospheric forcing (wind temperature, humidity, incoming solar radiation) and the heat conduction through the ice to compute the surface temperature. Furthermore, this energy balance contains parameterisations of incoming and outgoing longwave radiation and sensible and latent heat fluxes.
2. A model describes the heat conduction through the ice for given surface temperature and oceanic heat flux into the ice. The difference between the conductive and the oceanic heat fluxes determines the freezing (or bottom melting) rate. The surface melting is calculated if the surface temperature rises above 0°C. Then the surface temperature is kept at zero and the net energy is used to melt snow and ice.
3. A momentum balance including acceleration, Coriolis force, atmospheric and oceanic stresses, sea surface tilt and internal ice stress, from which the ice velocity is calculated. The internal ice stress is determined from a particular rheology which relates the internal stress to the deformation rate. A widely-used approach is based on the viscous-plastic rheology first introduced for sea ice by Hibler (1979).
4. Balance equations for snow and ice mass and ice concentration. These balance equations use the sea ice velocity and the freezing/melting rate determined from the heat conduction model together with the surface energy balance for ice and open water to calculate the new ice and snow thickness and ice concentration in each grid cell of the model domain.

The momentum balance (Second Newtonian Law), written in cartesian coordinates is given by

$$m \frac{D\mathrm{u}}{Dt} = -mf\mathrm{k} \times \mathrm{u} + T_a + T_w - mg\nabla H + \mathrm{F} \tag{7.1}$$

where u is the sea ice velocity, m is the sea ice mass per unit area, $f = 2\Omega \sin \phi$ is the Coriolis parameter (with the angular velocity of Earth's rotation Ω and latitude ϕ) and g is the gravitational acceleration. Eq. (7.1) includes inertial terms, Coriolis force, wind (T_a) and water (T_w) stresses, sea surface tilt H, and internal ice interactions **F** due to floe collisions and deformation (rafting and pressure ridge building)

According to continuum mechanics the internal ice forces within the sea ice pack can be written in terms of the divergence of the two-dimensional stress tensor

$\mathbf{F} = \nabla \cdot \sigma$, where a specific constitutive law has to be defined which describes σ as a function of the deformation (ϵ) or the deformation rate ($\dot{\epsilon}$), i.e. $\sigma = \sigma(\epsilon, \dot{\epsilon}, h_i, A)$. If the stress depends only on the deformation rate (which is given by the horizontal gradients of the sea ice velocity field), then the problem can be solved with the usual Eulerian approach applied in atmosphere and ocean modelling. In the Hibler (1979) model, the sea ice is considered to follow a nonlinear viscous-plastic constitutive law.

Local changes of snow and ice thickness (h_s, h_i) and compactness (A) are computed from continuity equations for these variables.

$$\dot{h}_i = -\nabla \cdot (\mathrm{u}h_i) + S_h \tag{7.2}$$
$$\dot{A} = -\nabla \cdot (\mathrm{u}A) + S_A \tag{7.3}$$
$$\dot{h}_s = -\nabla \cdot (\mathrm{u}h_s) + S_s \tag{7.4}$$

where $A \leqslant 1$. The first terms on the right hand sides describe the changes of the sea ice variables due to ice advection. S_h and S_A are thermodynamic growth terms, which are determined from conductive heat fluxes through the ice arising from conservation of energy at the upper surface of the snow/ice system. The S_h term is the sum of the ice growth over both open water and that part of the grid cell covered by ice. The growth rates are highly nonlinear functions of ice thickness, ice compactness and of atmospheric variables like temperature, wind, cloud cover and precipitation. The S_s term accounts for snowfall and melting of snow.

The model equations described above are solved numerically on a rotated spherical grid for the Arctic domain. The atmospheric forcing is taken from the NCAR/NCEP Reanalyses for 1958 to 1997 (for details see Hilmer et al., 1998). One of the main achievements recently was the determination of the optimal sea ice rheology for use in climate models (Lemke et al., 1997; Kreyscher et al.,1997; Harder et al., 1998; Kreyscher et al., 2000). Results of these investigations show that an optimised viscous–plastic rheology yields the best agreement between modelled and observed sea ice motion.

7.4 CONSEQUENCES OF THE MOVING SEA ICE COVER

The mean sea ice motion pattern for the period 1958–1997 is displayed in Fig. 7.5 (this figure appears in the full-colour section). It is described by a pronounced Beaufort Gyre and the Transpolar Driftstream. In most areas in the central Arctic and especially on the Siberian shelves a divergent drift pattern prevails. As a consequence more sea ice is frozen in these regions in winter than melted in summer, i.e. the net freezing rate is positive, in some areas more than 2 m/year.

As a result of the divergent drift pattern, ice is exported into the Nordic Seas where it is melted, producing a blanket of freshwater that stabilises the oceanic stratification. These freshwater fluxes are partly balanced by the inflow of salty surface water from the northward-flowing extensions of the Gulf Stream. Observations show that deep convection in the Greenland Sea is not occurring every year. It seems that deep convection is partly associated with weak sea ice export from the Arctic Ocean.

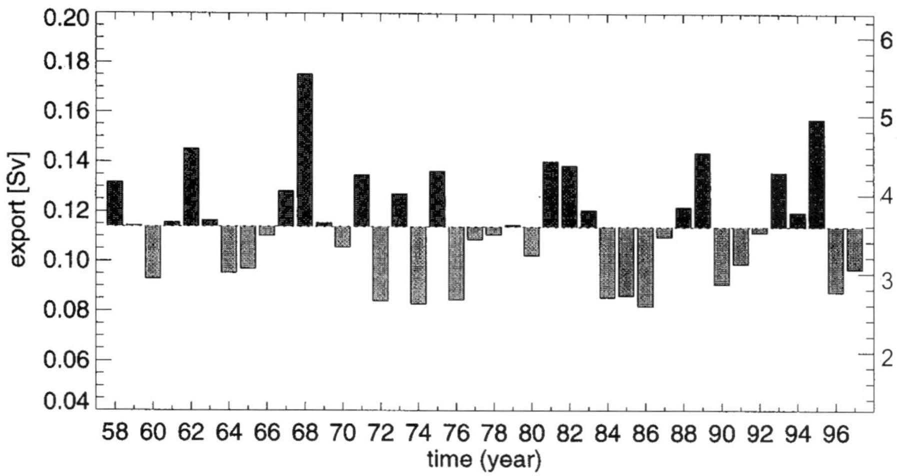

Fig. 7.6. Simulated variability of sea ice volume transport into the Greenland–Iceland–Norwegian Seas for the period 1958–1997 (courtesy M. Hilmer, IfM, Kiel).

The sea ice export into the Nordic Seas exhibits a large interannual variability which is displayed in Fig. 7.6. The most pronounced ice export occurred in 1968. This is the year when the Large Salinity Anomaly was first observed as a low salinity patch in the North Atlantic. The mean annual ice export into the Nordic seas amounts to 3600 km^3/year which is of the order of the river outflow of the Amazon.

Together with the improved analyses of operational weather forecast centres which provide realistic atmospheric forcing data, sea ice models are well capable of capturing the influence of the synoptic weather patterns on the sea ice motion. Figures 7.7 and 7.8 display the modelled ice motion for 13–16 and 17–20 January 1994, respectively. A comparison with Figs 7.2 and 7.3 reveals the close agreement.

7.5 TRANSPORT OF SEDIMENTS

Along with the sea ice, other incorporated material such as pollutants and, more importantly, sediments entrained during the freezing period is transported through the drift routes and is released during the melting period further down the Transpolar Drift Stream. During the course of time (generally a few years) sediments are transported from the source regions, mainly the Siberian shelf seas, across the Arctic Ocean. During summer or when the ice floes reach the ice edge, ice melting occurs and sediments are released into the water column contributing to the sedimentation on the sea floor. Due to the fact that sea ice sediments from different source areas have distinct mineralogical composition, it is possible to reconstruct the sea ice circulation from geological samples (see Eicken et al., 2000). In the following a numerical experiment is presented which investigates the connection between the source regions and release regions of sea ice sediments for the period 1979 to 1995.

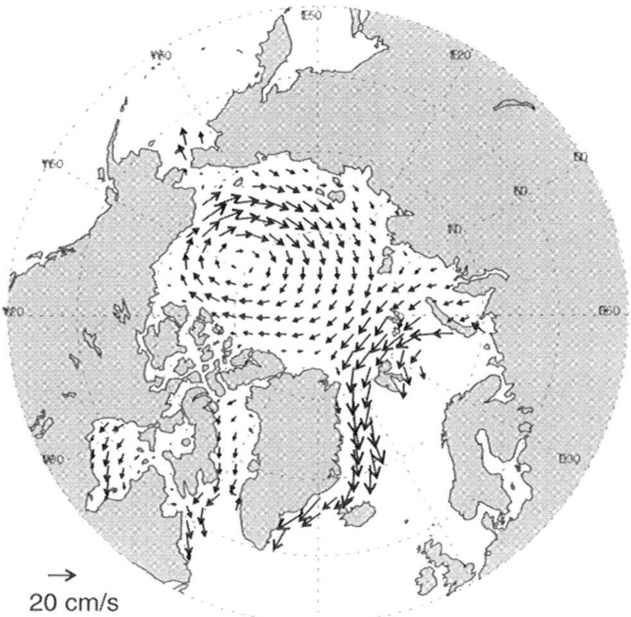

Fig. 7.7. Modelled sea ice motion for 13–16 January 1994 (courtesy M. Hilmer , IfM, Kiel).

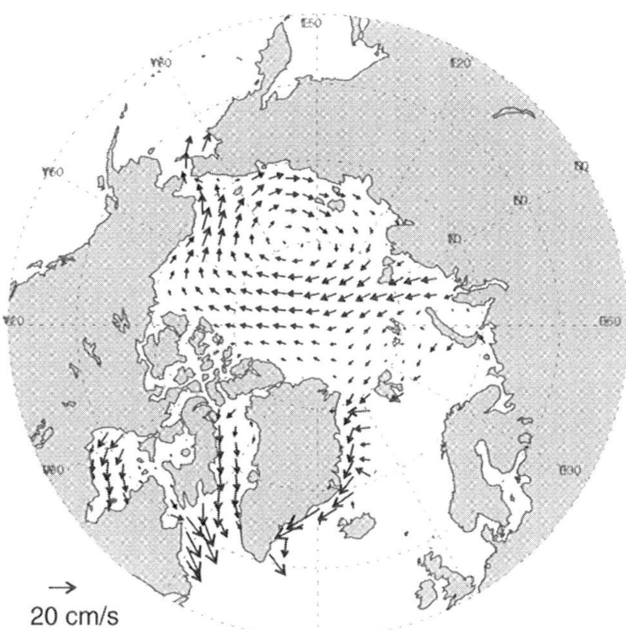

Fig. 7.8. Modelled sea ice motion for 17–20 January 1994 (courtesy M. Hilmer , IfM, Kiel).

This will give an estimate of the influence of the natural variability of the sea ice drift on the distribution of sea ice sediments in the Arctic Ocean and the adjacent seas.

For the simulation of the drift trajectories and the release regions of sea ice sediments the results of the optimised thermodynamic–dynamic sea ice model described above forced with atmospheric observations for the period 1979 to 1995 were used. From each of the four source regions (Kara Sea, western Laptev Sea, eastern Laptev Sea, western East-Siberian Sea) a set of 25 virtual sediment buoys were started at 79N (blue line in Fig. 7.9 – see colour section) for every November (freeze-up period) of each year of the simulation.

The trajectories of these buoys were calculated subsequently for a time period of three years. When the sea ice model calculates melting of the ice, sediment is released from the ice, and the geographical location of the sediment release is stored.

The results show a significant dependence of the area of sediment release on the source region (Fig. 7.9). Most of the sea ice sediments originating from the Kara Sea is released near the source region. The main release region is in the northern part of the Barents Sea and around Franz-Josef-Land (Fig. 7.9a). Sea ice sediments from the eastern and western part of the Laptev Sea reach the Barents Sea only with a very small probability (Fig. 7.9b, c). The main release region is the area of the Transpolar Drift Stream and the East Greenland Sea. Sediments from the eastern Laptev Sea also reach the area north of Greenland. Sediments originating from the western part of the East Siberian Sea are released in the central Arctic Ocean and in the East Greenland Sea. A significant release of sediments also takes place in the area north of the Canadian Archipelago (Fig. 7.9d).

In this numerical experiment sediments are released at a constant rate whenever ice melting occurs. This simplified parameterisation may overestimate the sediment release in the Central Arctic Ocean. A more sophisticated scheme which depends on sediment concentration and melting rates will most likely lead to less sediment release in the Central Arctic, where melt rates are usually low, and higher sediment release in the vicinity of the ice edges, where melt rates are generally larger. These improvements of the sediment transport model are the focus of current research.

8

Summary: ice rafting, ocean circulation, and climate change

During the Tertiary the Earth has undergone a long term climatic decline which led to the formation of the huge Antarctic Ice Sheet and later to the formation of sea ice cover in the Arctic Ocean and ice sheets of local and regional extent in the circum-Arctic lands. After the first appearance of coarse ice rafted debris in Arctic Ocean sediments, the ice rafted signal (the intensity of ice rafting) underwent several stages of intensification. After 1.6 Ma, the northern hemisphere has experienced more or less regular intervals of widespread continental glaciation, the ice ages, and interglacials with less ice similar to the present climate.

North of Ellesmere Island, 13 glaciations have imprinted themselves as coarse IRD peaks in deep-sea sediments during the last 1.5 Ma. The icebergs that dropped the IRD came from ice sheets in the northern Queen Elizabeth Islands and drifted on an eastward route towards Fram Strait and further into the Greenland Sea. Except for a rather small area north of Ellesmere Island, the Arctic Ocean in general has been affected by several incarnations of three large ice sheets: the Laurentide Ice Sheet (LIS), the Innuitian Ice Sheet (IIS), and the Barents-Kara-Svalbard Ice Sheet. Icebergs that calved off the LIS carried a mixture of carbonates, clastic sedimentary and crystalline rocks from Victoria Island and adjacent parts of Arctic Canada across a wide arch through the Arctic Ocean. This carbonate-mode of the Arctic circulation was powered by a strong atmospheric pressure gradient between the LIS and Fram Strait. The carbonate-mode was not just a thin sheet of ice floating on top of inert waters but a surface current of at least 400 m thickness. It dominated during extended periods of time and was capable of emptying the western Arctic Ocean of most if not all its microorganisms, thereby creating abiotic conditions. In the carbonate-mode, the Arctic Ocean most vigorously exchanges waters and ice with the sub-Arctic seas. The atmospherically driven surface outflow from the Arctic Ocean into the Greenland Sea amounts to a volume flow of 9 Sverdrup and was compensated by an inflow of waters at some sub-surface level.

Icebergs that calved off the western IIS (Ellef Ringnes Island and its surroundings) carried a load of quartz grains and clastic sedimentary rocks, while those from the northern IIS (Ellesmere and Axel Heiberg islands) carried crystalline rocks, carbonates, clastics and small amounts of quartz. The icebergs mostly drifted northeastward along the Canadian Arctic continental margin. Like icebergs from the LIS, icebergs from the IIS left the Arctic Ocean on a relatively direct route.

Icebergs from the northern Eurasian ice sheets (Svalbard, Barents and Kara ice sheets) carried a load of primarily quartz, some clastics and crystalline rocks. The icebergs took a westward drift in the carbonate-mode, leaving the Arctic Ocean directly through Fram Strait. This drift direction is not very different from the Transpolar Drift of today. But in the quartz-mode, the icebergs from northern Eurasia drifted in a counterclockwise circulation and could periodically enter the western Arctic Ocean and influence the IRD composition there.

Both modes of the Arctic circulation were driven by strong atmospheric pressure gradients. In the quartz-mode, the Barents and Kara ice sheets produced a large, stable, high-pressure system which pushed ice towards the central Arctic Ocean. There, the flow encountered somewhat weaker winds from the American side which made the flow turn towards Fram Strait. The quartz-mode provided a direct path for moisture-laden air from the North Pacific to the high Canadian Arctic, where the moisture fuelled glacier growth. As the ice sheets in the Canadian Arctic grew in regional extent and thickness, an area of high atmospheric pressure developed over them and gradually became the dominant driving force of the Arctic circulation, changing it into the carbonate-mode. While the quartz-mode is a secluded circulation with limited interaction with the sub-Arctic seas, the carbonate-mode is strongly interactive and at times could have flown at a rate of 9 Sverdrup.

The flow of ice from the Arctic to sub-Arctic seas had consequences for the watermass stratification there. When large amounts of ice were exported, they melted in the sub-Arctic seas and reduced both the surface salinity and temperature of the approaching North Atlantic Current. This current therefore became specifically lighter and continued its northward passage as a cold, iceberg carrying surface current in the Norwegian Greenland Sea. In exceptional circumstances this current could reach as far north as Fram Strait, in which case I like to call it the 'long loop'. When less ice was advected into the sub-Arctic seas, the North Atlantic Current cooled less quickly and remained relatively high saline. It entered the Norwegian Sea at its southeastern entrance and flowed northward as a Proto-Norwegian Current and kept the southeastern Norwegian Sea partly ice free. At about 67° N it became heavier than the glacial waters of the NGS and descended beneath them (the short loop).

These processes triggered intricate interactions, complicated oscillatory effects in which one current system creates the other and vice versa. When the carbonate-mode was in full effect, large amounts of ice drifted into the northwestern North Atlantic, melted, and were assimilated into the Pleistocene North Atlantic Current. This current continued to flow north through the NGS and in some instances even entered the eastern Arctic Ocean through eastern Fram Strait. This process created the long loop and, if completed in the eastern Arctic Ocean, changed the Arctic surface circulation from the carbonate- into the quartz-mode. Conversely, whenever the quartz-mode was in full effect, the amounts of ice that were available in the NW

North Atlantic to lower the salinity and temperature of the North Atlantic Current were less and the current remained higher saline and warmer. Its descent to subsurface depths in the Norwegian Sea and continued northward flow (much like the West Spitsbergen Current which continues in the eastern Arctic Ocean as the 'Atlantic-layer' at the present) then provided the kinetic energy for a strong outflow of Arctic surface waters across the entire width of Fram Strait, the carbonate-mode. Therefore, via its diminished southward ice volume flow, the quartz-mode triggers the carbonate-mode and vice versa.

The feedback chain worked (and works?) as follows:

Step 1: The carbonate-mode at full strength supports the long loop in the NGS.
Step 2: The long loop gives rise to the quartz-mode circulation in the eastern Arctic Ocean.
Step 3: The quartz-mode produces the short loop.
Step 4: The short loop triggers the carbonate-mode of the Arctic circulation.

The primary control on this feedback chain is the amount of freshwater equivalent (sea ice, icebergs, glacial meltwater) in the northwestern North Atlantic, which in turn depends on the iceberg calving rates of the Pleistocene ice sheets and the surface circulation. In order to evaluate what role the atmospheric circulation played for the feedback loop, let us first consider the consequences of an unobstructed feedback loop without atmospheric forcing and constant, steady iceberg calving rates. In that case the feedback chain would have operated continually, going from step 1 to step 4, then back to 1 and so forth. The different states of the resulting surface circulation would have changed like a pendulum. Consequently, in the Arctic Ocean the carbonate- and quartz-modes would have switched back and forth, while in the Norwegian Sea the long loop alternated with the short loop.

The resulting IRD composition in the Arctic Ocean should be a thorough mix of lithologies from both the Eurasian and the American sides and, since the pendulum operated all the time, the IRD composition should have stayed the same through time. Because this was not observed, the Arctic Ocean component of the surface ocean circulation was strongly controlled by the atmospheric pressure distribution, which in turn was proportional to the location and size of the large Pleistocene ice sheets over North America and northern Eurasia. When the Barents–Kara Ice Sheet was at its largest geographic extent, during middle Weichselian times, its high pressure system drove the quartz-mode. Later, the Laurentide and Innuitian Ice Sheets grew to maximum extent, and the zone of highest atmospheric pressure shifted to North America, driving the carbonate-mode.

But the presence of Fe oxide grains from both sides of the Arctic Ocean shows that the two dominant current systems oscillated on time scales shorter than those represented by the sediment samples. Therefore the atmospheric circulation controlled the long-term surface circulation, which was modified on shorter time scales by the density-driven feedback chain.

In the Norwegian Sea, however, the observed IRD composition lends strong support to the density control. Many samples from the eastern Norwegian Sea contain IRD from sources reaching from the Baltic region (southern Sweden and

Denmark) up along the Norwegian coast to Spitsbergen. This supports the notion that north- to northwestward currents (the long loop) alternated with south- to southwestward currents (the short loop), and therefore the feedback chain was very influential for the circulation of the Norwegian Sea.

The Arctic Ocean surface circulation is driven by the atmosphere, and density differences of the watermasses play a minor, modifying role. In the Norwegian Greenland Sea, the roles are reversed: density differences of the ocean are primarily responsible for the circulation and the atmosphere plays a minor role.

8.1 PLEISTOCENE ICE SHEET DYNAMICS

The Weichselian ice sheets began to grow in the farthest northern lands on Earth. Triggered by dropping temperatures as a consequence of changing orbital parameters and fed by moisture from the open Norwegian Sea, glaciers began to grow in Svalbard and in northern Norway. At some time between 120 and 100 ka, the ice sheets reached sea level and the first icebergs from the north drifted southward through the Norwegian Sea. Amplified by an increasing albedo, the ice sheets grew and coalesced over the Barents Sea and expanded over Scandinavia. Gradually the clastic sedimentary rock-bearing icebergs from the north were replaced by icebergs from the western flank of the Fennoscandian Ice Sheet with abundant crystalline rocks. During the middle Weichselian, ice sheet growth produced a large northern Eurasian ice sheet that covered, in addition to the Barents Sea and Scandinavia, also the Kara Sea. Ice sheet growth in North America lagged behind Eurasia. During the time of maximum cold, from 23–18 ka, different parts of the western Fennoscandian ice sheet produced icebergs in distinct pulses. These icebergs carried a load of low diverse crystalline rocks.

Immediately after the last glacial maximum, the Barents Ice Sheet disintegrated in several distinct stages. First a large mass of ice was lost from the western Barents shelf. The loss led to a separation of the once contiguous Barents Ice Sheet into a northern part over Svalbard and parts of the adjacent Barents Sea, and a southern part that covered northern Norway and parts of its shelf. Between 18 and circa 9 ka, all of the Weichselian glaciers over Scandinavia were lost and modern climatic conditions were established. The ice sheets of northern North America reached their maximum extent only after 30 ka, after the Kara Ice Sheet had largely vanished. The far eastern part of Siberia was not glaciated during the Weichselian.

It appears that ice sheet formation is initially triggered by an orbitally induced drop of the global temperature, most probably sometime during the precessional 23 kyr cycle. Dropping summer temperatures are the most important factor. Once initiated, earth-borne effects such as increasing albedo and changing moisture sources take over as the most important factors. For instance, the initial ice growth was strongest over northern Eurasia, progressing from the west (Svalbard) to the east (Kara Sea, Zevernaya Zemlya), with the moisture coming from the North Atlantic. When the Kara Ice Sheet was large and the quartz-mode of the Arctic circulation operated, from sometime before 50 ka to approximately 38 (?) ka, a formidable moisture source existed in the North Pacific, feeding ice sheet growth in Arctic North America. Once the ice sheets reached a critical minimum size, they formed large, high-pressure cells which controlled the ice drift and ocean circulation. As a side effect, the atmospheric high pressure reduced the amount of moisture and therefore slowed ice sheet growth.

The amounts of icebergs that were discharged during any given time period from the Late Pleistocene northern hemisphere ice sheets were partly related and in part unrelated to climate changes. While the northeastern Laurentide Ice Sheet produced iceberg swarms synchronous to temperature changes over the Greenland Ice Sheet (= related to climate), the western Fennoscandian Ice Sheet had five distinct iceberg pulses from different sections during the last glacial maximum when the marine oxygen isotope values show no signs of a changing climate at all (= unrelated to climate). Therefore, in addition to climate changes, other physical factors such as the vertical ice sheet profile, the snow accumulation rates, and the conditions at the glacier bed (the friction between glacier ice and hard crystalline rocks versus softer clastic rocks versus unconsolidated marine shelf sediments) control the ice flow velocity and thus the iceberg calving rates.

8.2 HOLOCENE CLIMATE VARIATION

The IRD record of the Norwegian Sea demonstrates that the Holocene was not a 9000-year-long period of steady, stable climate conditions. In contrast to the marine oxygen isotope record which shows flat values throughout, the IRD densities and composition have changed considerably. Icebergs and sea ice from sources reaching from Svalbard to the East Siberian Sea have drifted southward through the Norwegian Sea, opposite to the prevailing currents of today. The southward currents that carried the ice were most likely extensions of the East Spitsbergen Current. This current intruded southward between two northward-moving currents, the Norwegian Current and the Norwegian Coastal Current. Such southward expansions were frequent and occurred during the time span represented by every analysed sample, within less than 1000 years. These excursions represent times of a weakened Norwegian Current. Whether strong northerly wind over the Arctic domain triggered a southward surface flow with ice which then, in turn, weakened the Norwegian Current, or whether it was a weakened Norwegian Current that allowed the southward expansion of ice cannot be answered.

The instability of the Norwegian Current also has not been the same throughout the Holocene. Initially, during the early Holocene, ice rafting was intense and the amounts of IRD in the sediments high. The Norwegian Current then gradually became stronger and was at its strongest during the middle Holocene at around 4.3 ka when ice rafting was minimal and most likely the warmest conditions during the Holocene were obtained. At this time, changes in the atmospheric circulation also occurred in other places on Earth (Akkadian Empire, Northeast Syria, the Arabian Sea, off the coast of Chile), which suggest that the shift in the Norwegian Sea amounts to a global rearrangement of the atmospheric pressure distribution. Not only the intensity of ice rafting, but also the directions from where the ice came changed at 4.3 ka. The shift, ironically, must have created enjoyable climate conditions in Northwest Europe, whereas people in Northeast Syria were forced away from their land by the initiation of very dry, desert-like conditions.

After this middle Holocene period of relative warmth in the Norwegian Sea, ice rafting became again more frequent. In fact, if we take the grain densities as a direct measure of the frequency of ice rafting, ice rafting became increasingly frequent towards the present. Therefore, and because the IRD grain densities at the present are as high as directly after the last ice age, the global climate may well be in the initial stages of another ice age.

9

Major findings

1. Ice rafting is a physical process that not only monitors the surface currents but also actively influences the climate: massive quantities of ice drifting from one place to another lower the salinity of the surface waters wherever the ice melts. The amount of freshwater absorbed into the surface current can make the difference between an ice-covered cold surface current or an ice-free, more temperate subsurface current.

2. The trace of past ice drift directions is always recognisable in the composition of the IRD that was dropped by the icebergs and by sea ice. Pronounced differences in the petrographic composition of different iceberg sources assist in the identification.

3. Sea ice rafting carries mostly fine-grained sediment. Sea ice IRD has a smaller maximum grain size (of about 2 mm), low grain concentrations of 0–20 grains $>250\,\mu m$ per one gram sediment, and is low diverse.

4. Iceberg rafted debris, by contrast, is much coarser with largest grains up to boulder size and grain concentrations generally in the 100–300 grains/g sediment range. The diversity is much greater than for sea ice rafted debris.

5. IRD in the Arctic Ocean consists of three basic types; (1) carbonate-dominated, (2) quartz-dominated, and (3) crystalline rock-dominated. Type 1 IRD can be found in the western and central Arctic Ocean (AO), type 2 IRD in the central and eastern AO, and type 3 IRD only in a sector from north of Ellesmere Island to Fram Strait.

6. Type 1 IRD was transported by icebergs from the northwestern Laurentide Ice Sheet which covered extensive Paleozoic carbonate-terrains in the southwestern Canadian Arctic Archipelago.

7. Type 2 IRD has two places of origin: it came from the Kara Ice Sheet during middle Wisconsinan times and from the Innuitian Ice Sheet during older glaciations, perhaps at times from both.

8. Type 3 IRD came from the northern Innuitian Ice Sheet and consists of lithologies that typify the Pearya Terrane of northern Ellesmere Island.

9. In the central Arctic Ocean, carbonate IRD alternated with quartz IRD repeatedly during the last 1 Ma.

10. In the western Arctic Ocean, carbonate-dominated IRD with clastic rocks and crystallines from the Canadian Shield was dominant throughout.

11. North of Ellesmere Island, crystalline IRD alternated with quartz IRD.

12. The dominating ice drift direction in the Arctic Ocean was from the margin of the large Laurentide and Innuitian ice sheets in a wide curvilinear path across the pole towards Fram Strait. In this mode, the carbonate-mode of the Arctic circulation, the Arctic Ocean exported massive quantities of glacial icebergs and sea ice into the Norwegian Greenland Sea.

13. The carbonate-mode alternated with the quartz-mode, which was a gyral, counterclockwise circulation in the eastern and central Arctic Ocean.

14. During the Pleistocene Glaciations, the circulation of the Arctic Ocean, the Greenland Iceland Norwegian (GIN) seas, and the North Atlantic was connected to form a long north to south loop: waters entered the GIN seas through the southeast and flowed north along the northwestern European continental margin. The ice-laden waters entered the Arctic Ocean through the eastern Fram Strait, circulated anticlockwise with the quartz-mode and returned through western Fram Strait, flowing south along the East Greenland continental margin.

15. For most of the time, however, the waters of the Pleistocene North Atlantic Current were more saline and heavier and sank to a sub-surface level in the Norwegian Sea. This descent opened the way for generally southward to southwestward flow in the northern Norwegian Greenland Sea.

16. During most of the Pleistocene, the ice drift and surface water motion in the Norwegian Sea switched between northward, westward and southward. In the Greenland and Iceland seas, the flow was generally towards the south. Between the north- to northwestward flow in the Norwegian Sea and the southward flow in the Greenland Sea, an anticlockwise gyre developed in which icebergs from sources several 1000 kilometres apart were mixed together.

17. The Barents Ice Sheet existed in almost the same configuration five times during the Pleistocene Glaciations, during oxygen isotope stages 12, 10, 8, 6 and 4–2.

18. The Innuitian Ice Sheet in the northernmost part of North America existed 13 times during the last 1.5 Ma.

19. The Laurentide Ice Sheet existed at least six times during the last 1 Ma.

20. Ice sheet growth in Eurasia during the Weichselian Glaciation began in the northernmost lands close to the Arctic Ocean. Glaciers grew on Svalbard, extended onto the Barents shelf, and coalesced with ice from northern Norway to form a contiguous ice mass over Scandinavia, the Barents Sea and Svalbard.

21. Iceberg calving from this ice sheet occurred in distinct pulses which were unrelated to climate changes.

22. The final disintegration of the Pleistocene ice sheets began shortly after the last glacial maximum, at 17.5 ka. The first part of the ice sheets that disappeared was the southwestern Barents Ice Sheet. Icebergs drifted southward through the Norwegian Sea. As this southward drift continued, northward-moving currents established themselves to the east and west of the southward flow. Over the course of the following approximately 6000 years, these northward currents gained strength and gradually developed into the present-day circulation.

23. The Holocene warm period began approximately 9000 years before the present. The Holocene climate is not stable but oscillates between the dominant warm, ice-free scenario of today and times when ice and cold surface waters moved from the Barents Sea southward through the Norwegian Sea (Holocene cold excursions). The ice and icebergs that drifted in these cold waters came from Svalbard and places in northern Siberia.

24. The Holocene cold excursions were most likely of short duration and caused by a rearrangement of the atmospheric pressure distribution over the Arctic domain.

25. In addition to these shorter changes there is a longer trend through the Holocene from initially frequent cold excursions to very few in the Middle Holocene. This time period was the warmest and was characterised by a very strong and wide Norwegian Current, which at this time kept the Norwegian Sea truly ice-free. Thereafter, the IRD record shows an intensification of the cold excursions that is increasing towards the present.

26. The Middle Holocene at approximately 4200 years bp was a time during which a global climate change occurred: the surface currents shifted in the Norwegian Sea, and changes of the atmospheric circulation occurred in places such as Arabia and off the coast of Chile.

27. These climate changes on different time scales as well as the dynamics of continental glacier growth are governed by changes of the atmospheric pressure distribution over the Arctic domain, and these changes, in turn, are global.

28. The most important element of the climate in the high northern hemisphere, in the view of this author, is the location, size and persistence of high pressure in the circum-Arctic: as ice formed initially over northwestern Europe, the ice sheet over the Barents and Kara seas produced a high-pressure system which triggered the quartz-mode of the Arctic circulation. This mode opened the way for moisture from the Pacific to the Canadian Arctic islands, where the Laurentide and Innuitian ice sheets began to expand. As they reached a certain minimum size while the Kara Ice Sheet shrank, the high pressure cell shifted to northern North America and henceforth drove the carbonate-mode of the Arctic circulation.

29. It appears that one of the most fundamental elements of earth-borne climate changes is the anticlockwise motion of a climate signal around the periphery of the

Arctic Ocean (in the direction of the Earth's rotation). The climate signal can be the direction of ice drift, the location of atmospheric high pressure, or the area where glaciers grow. Glacier growth, for instance, commenced in northwestern Europe, then moved eastwards around the Arctic Ocean to the Kara Sea and North America. Similarly, the direction of ice drift changed from the Kara Sea during the middle Weichselian to the East Siberian Shelf (fine-grained interval in the central Arctic box cores between the sediment sections dominated by quartz and carbonate IRD), and finally to northern North America.

30. The shift of the climate signal in the direction of Earth's rotation occurred on time scales of thousands of years and on time scales much shorter than that, perhaps as short as several decades.

10

Some afterthoughts

In the ongoing debate about climate change, mostly understood as anthropogenically influenced global warming, some details, which in my opinion are very important, have been swept under the carpet. Let us take, for example, the global average temperature curve, for the last 150 years. Everybody who reads this book has seen this curve, which is often taken as evidence that global warming is going on. The curve shows a trend to slightly warmer temperatures at present than 150 years ago, before the beginning of industrialisation. But the question that nobody asks is what the database of this temperature curve is. The first reason why any such curve inevitably must be skewed is because weather stations are not evenly distributed across the globe. For most of the oceans, which constitute 70% of the surface of the Earth, there simply aren't any measurements available except perhaps for a few shipborne observations. Furthermore, most weather stations have experienced significant changes in the geographical setting in which they are located. For instance, 100 or 150 years ago, a particular weather station may have been in a forest or a park-like environment. But with the population explosion on Earth, cities grew in geographic extent and certainly began to enclose some of these weather stations. At one point, the station may no longer be in a green park but rather surrounded by buildings, parking lots and shopping malls. This urbanisation effect leads to artificially higher temperature measurements which are impossible to filter out from the natural signal. One of the meteorologists on board 'Polarstern' in the summer of 1989 mentioned that, if all weather stations that became part of an urban environment were eliminated from the temperature database, the rising trend disappears and the temperatures (in the countryside) stay essentially at a flat, unchanged level.

The reader should understand that in science, as in other sectors of public life, the outcome of a study is very often guided, if not determined, by an *a priori* idea, a tenet. One could also call it a belief. In the case of global warming, this belief was that, if enormous amounts of greenhouse gases are released into the atmosphere, a temperature rise *must* occur. This prior assumption has guided scientific thinking and triggered a true deluge of investigations all desperately trying to prove just that. But climatologists are not the only ones who fall into these traps of preconceived ideas. Here is a loose sampling of some recent scientific blunders:

- Cold nuclear fusion, proclaimed loudly by two chemists in the late 1980s as the ultimate answer to mankind's energy demands, has not been accomplished once.
- The universe was once believed to have an 'age' of 8 billion years. All it took to pulverise this idea was to send out the Hubble telescope; the age is now believed to be 16 billion years. Who knows what it will be next ...
- The 'discovery' of 'life on Mars' was boldly announced a couple of years ago. Suffice it to say that it has become very quiet on this subject.
- And more recently, NASA scientists proclaimed at a press conference (why bother with the peer-review process?) the first discovery of extrasolar planets. Needless to say, this 'discovery' was retracted only two days later. The planets turned out to be faint stars.

It's amazing what results stiff competition for research grants and the pressure to publish can produce.

In the case of the climate debate, what has been totally forgotten is the fact that natural climate changes occur as well as man-made ones, and on time scales shorter than 150 years. And the only way to detect these changes is from the geologic record of marine sediments.

Therefore, this book was written to present some of these naturally-occurring climate changes. I tried to remain as honest as possible and took the conclusions about the changing surface current directions as far as the IRD record permits. Only with regard to the driving forces did I resort to some speculation, but even these speculations should be within realistic limits. The most important result is that circulation changes that imprinted themselves over the course of hundreds and thousands of years in the IRD record also occur on shorter time scales.

At the present, we cannot resolve these changes on multi decadal or shorter time scales relevant to humans. This is simply because the low, and in some cases very low, sedimentation rates of the polar oceans do not permit time resolution at these short scales. But recent progress in the analysis of Arctic Ocean sediments has shown that it is possible to find areas with high resolution. This, and the prospect of a new polar icebreaker for more than 200 days of use per year, hold the promise that the mystery of the driving forces of climate change may be solved eventually. This book describes some small but decisive steps along the way.

Endnotes

1 El Niño is the name for an occasionally-occurring reversal of the circulation in the equatorial Pacific. Normally, during periods of 5–10 (or more) years, the prevailing cold currents flow from the south up along the west coast of South America, bend westwards near the Equator, and then flow to the western equatorial Pacific. Under this scenario, upwelling conditions exist off the coast of Peru, bringing cold, nutrient-rich deep-waters to the surface and maintaining a rich marine fauna and flora. But from time to time, the usually stable atmospheric high pressure zone over the western equatorial Pacific breaks down and, as a result, the sea surface slope is tilted towards South America. In this El Niño situation, warm, nutrient-depleted surface waters flow towards the Peruvian coast and remove the food base for most marine organisms which, as a consequence, die. The El Niño situation is commonly believed to cause widespread disruptions of the climate elsewhere on Earth, and thus be more than a regional phenomenon.

2 Global Climate Models (GCMs) have been made possible as a result of the advent of powerful computers. The models are run by exerting the principal physical forces that drive the climate (wind stress, density structure of the ocean, thermohaline circulation, and the relative sea surface height) onto a global ocean which is represented by a grid of cubes of variable size and multiple layers for the ocean (currently about 21 for the simulation of the global ocean). The model domain can be adjusted to any area of interest, and to the capabilities of the available computers. Model runs can be performed to simulate the atmospheric or the ocean circulation. Since their invention, climate models have become more sophisticated and are now capable of simulating the present-day climate very closely. They have thus also been used to predict the consequences of increasing CO_2 levels in the atmosphere.

3 The SHEBA project (acronym for Surface Heat Budget of the Arctic Ocean) was designed to obtain synoptic (real-time) observations of important physical, chemical and biological parameters continuously over the course of a whole year. To do this, the Canadian icebreaker Des Grosseliers was moved to a position in the southern Beaufort Sea in September 1997 and allowed to be

frozen into the pack ice at the beginning of winter. As the ship drifted with the ice pack, very much like Fridtjof Nansen's Fram did more than 100 years earlier, numerous measurements were made from instruments spread around the ship. This experiment has yielded unprecedented high-quality data from the Arctic Ocean. Data evaluation is in progress, but by the time of publication of this book many of the SHEBA results should be published. All those who absolutely cannot wait are referred to the SHEBA web site at http://sheba.apl.washington.edu/images.

4 The Earth is surrounded by a network of satellites that monitor parameters as diverse as the extent and types of sea-ice (whether there is relatively thin and soft first-year ice or comparatively thick and rigid multi-year ice), microwave radiation, air temperature, chlorophyll content in the surface waters (as a measure of the phytoplankton abundance), ozone concentration in the upper atmosphere, ocean currents, sea surface height, etc. The satellites are parts of an Earth-observing system designed to detect signs of impending climate changes.

5 The North Atlantic Drift is the northern extension of the Gulf Stream system. Warm waters (28°C and more) from the Gulf of Mexico move around the southern tip of Florida northward, up along the east coast of the southeastern Atlantic states. Near Cape Hatteras, the current moves offshore and the waters move northeastward across the Atantic. Further east, one branch moves back to the south west of Africa to complete a clockwise circulation in the southern North Atlantic. In the north, the Gulf Stream changes its name to North Atlantic Drift and flows northwards along the northwestern European continental margin. The current gives off heat and moisture to the atmosphere and becomes gradually cooler. Upon entering the Arctic Ocean west of Spitsbergen, the waters have cooled to a temperature at which they become more dense than the Arctic surface waters. As a consequence, they sink and continue to flow into the Arctic Ocean at a water depth of more than 400 metres. The North Atlantic inflow is compensated by a surface outflow from the Arctic Ocean which becomes the East Greenland Current.

6 The Transpolar Drift (TPD) is one of the most pronounced features of the Arctic Ocean's circulation. It is a movement of pack ice from the Chukchi Sea, located north of the Bering Strait, northward to the North Pole and then towards Fram Strait and into the northern Greenland Sea. The motion is primarily driven by the atmospheric circulation which transfers its kinetic energy via friction to the ice surface. The TPD was discovered by Norwegian oceanographers (Sverdrup, Helland-Hansen, Nansen) in the nineteenth century and was believed to be so strong and reliable that Fridtjof Nansen entrusted the life of his men to it when he allowed his ship, the Fram (Norwegian for 'forwards'), to be frozen into the pack ice in 1893. He calculated that in three years the TPD would carry him across the North Pole and then into open water in the Greenland Sea. Nansen was basically right except that he missed the North Pole. He was a very courageous man indeed, because newer advances in paleoceanography (Bischof and Darby, 1999) indicate that it was equally likely he may have ended up in the Queen Elizabeth Islands, Canadian Arctic Archipelago.

7 The albedo is a measure of the amount of solar radiation that is reflected from the surface of the Earth. The air temperature at the Earth's surface depends on the amount of incoming solar radiation being converted into heat. Brightly coloured surfaces such as ice and snow reflect much of the incoming radiation and convert little into heat. On the contrary, dark, matt surfaces such as grey sandstones or black shales reflect little and convert most of the radiation into heat. For example, while air temperatures above the snow surface in Antarctica can be −10°C, the temperature measured directly underneath dark grey pebbles in the Dry Valley region can be +12°C. The higher the albedo, the more radiation is reflected and vice versa. The albedo is a dimensionless value between 0 (no reflection) and 1 (complete reflection). Dark surfaces have very low albedos, whereas snow and ice surfaces have albedos of up to 0.9. The strength of this effect can be felt by anyone who, without the help of sunglasses, has ever tried to brave the glare of snow on a bright, sunny day during skiing vacations!

8 Orbital forcing: the motion of the Earth around the Sun is neither circular nor constant. The Yugoslavian mathematician Milutin Milankovitch calculated that the Earth has a slightly elliptical orbit (eccentricity), that the plane of Earth's orbit is tilted at a low but variable angle against the plane of the solar system (obliquity), and that the Earth's rotational axis wanders around in circles, performing one whole revolution in about 23,000 years. The periodicities of obliquity are 41,000 years and for eccentricity 100,000 years. As a result of these changing orbital parameters, the Earth receives variable amounts of solar radiation, sometimes more and at other times less. These changes control the Earth's heat budget and are the main driving forces of the ice ages.

9 Postglacial isostatic rebound: the Earth's lithosphere, consisting of the Earth's crust and parts of the upper mantle, is normally in isostatic equilibrium with the underlying asthenosphere. The lithosphere, so to speak, floats on the asthenosphere very much like ice floats in water. The thicker the ice (lithosphere), the deeper it is submerged in the water (asthenosphere). During the Pleistocene glaciations, ice formed on various places of the northern hemisphere, and the crust began to sink into the mantle by an amount equivalent to the ice overload. When the ice melted, the crust was freed of the overload and began to rise. But because the squeezed-out mantle material is very viscous, the emergence was much slower than the ice recession. Postglacial emergence in some cases took 8000–9000 years after the end of the Pleistocene and is still going on in Sweden. Therefore, glacial till deposits left by the Pleistocene glaciers may have been below the sea level initially, but then emerged above the sea level as the rebound continued.

10 Sampling. The two coring devices used were the kastenlot (KAL) and the Grosskastengreifer (GKG). The KAL is a large gravity core with a square cross section of 30 cm edge length. This core consists of 6 m long segments that can be combined to form 12 or 18 m long barrels. The core is pushed into the sediment by a weight of 3 tonnes. This type of corer is open and does not possess a piston. For glaciomarine sediments such as those in the NGS, the maximum recovery is about 9 m. The GKG is a giant box grab consisting of a cubic metal box of 50

cm edge length and a shovel. After penetrating the sediment, the shovel closes underneath and a 0.25 cubic metre area of undisturbed sediment can be recovered. The advantage of these coring techniques over conventional small diameter piston and gravity cores is the recovery of large amounts of undisturbed sediment.

In the GKG samples, the primary objective was to separate the surface layer with very little IRD from the underlying glaciomarine sediments, and to split the very coarse layers in the glaciomarine unit from each other. Therefore, every multiple-centimetre thick pre-Holocene sample from our GKGs contains IRD from only one single dropstone-rich IRD layer. In those box cores which contained either very thick Holocene units, or Holocene sediment only (23067–23072), the Holocene section was split into as many as five units at randomly selected intervals. This sampling strategy produced large volume samples of 2–17 cm thickness over a surface area of 0.25 sq m. However, for the Poseidon samples (five digit sample numbers beginning with 16), a different sampling scheme was applied. Here, horizontal slices of 1–5 cm thickness were taken until the whole box core content was consumed. This yielded higher resolution samples than in the Meteor (23000-series) cores. Since no oxygen isotope stratigraphy was obtained for the Poseidon cores, the stratigraphy of the nearest Meteor core was transferred.

Sampling in the longer kastenlot cores focussed on the coarsest intervals. It is Kiel University's practice to process the large kastenlot cores directly during the cruise on which they were taken. The core is opened, and a fresh cross section is created using a thin, sharp metal wire. The core is then being described and sampled according to protocol. After other types of sampling consumed some of the sediment, the author was free to take large chunks of coarse sediment. Up to 20 cm-thick samples across sections of up to 30 × 30 cm were taken. The coarsest intervals were found by poking through the mud with a spatula or a knife. Only those sections where abundant ice rafted grains were identified – by the scratching sound they made upon contact with the spatula – were chosen for analysis. This method produced samples from the coarsest intervals. Therefore, our kastenlot samples, in general all those from core depths of 50 cm and below, represent periods of intensive ice rafting.

11 Petrographic IRD analysis. The samples were sieved through a 0.5 mm screen to retain the coarse grained particles. A random split of the >0.5 mm fraction was analysed quantitatively under a binocular microscope at 20 to 40x magnification. Selecting 0.5 mm as the minimum grain size accomplished two essential goals. First, it ensured that even the relatively fine grained Holocene sediments would still yield enough grains for meaningful statistical treatment; second, a minimum size of 0.5 mm proved to be sufficiently large to allow for accurate grain identification. It is an unwritten rule among mineralogists and petrologists that a piece of rock has to be at least the size of a (German) Mark coin to reveal all the features needed for proper identification. For all those who have never had any of this soon defunct currency, that means a piece about 3 cm in diameter. Pieces of that size were usually found abundantly in the Pleistocene

sections of our cores and used for comparison with the smaller fragments, and to verify the petrographic determinations that were made. In addition to this, microscopically-thin slices of rock fragments were made to corroborate the petrographic determinations made under the binocular. In total, 114 categories for rock and mineral fragments were created. For some of the largest samples, sub-splits of other grain size fractions were also analysed quantitatively. These size ranges are 2–4 mm, 4–6.3 mm, and >6.3 mm. In addition, many if not all of the larger rock fragments were individually described.

12 The sample is from 567–584 cm depth of kastenlot core 23058. The core is located in the Lofoten Basin, and the analysed sample is from a regular sediment sequence beneath a thick, black, sandy turbidite unit. Therefore, its age is not known.

13 The light coloured intrusive rocks (granites, syenites, diorites, tonalites, etc.) and gneisses in Norwegian Sea IRD assemblages contain more feldspar than quartz. Thus, if the quartz and feldspar grains were entirely derived from the decay of such rocks, feldspar grains should be more frequent than quartz, and the ratios should be lower than 1.

14 The core chronology is based on work done by Elke Vogelsang, Mara Weinelt, Michael Sarnthein, Antje Voelker and Rüdiger Henrich on sediment cores from the Norwegian and Greenland seas. These investigators established oxygen isotope stratigraphies (see footnote 14a) and ^{14}C AMS age determinations for the box cores and for the long kastenlot cores presented in this book (23059, 23062, 23068, 23071 and 23074). The ages of the major stage boundaries are taken from Berggren et al. (1995). To establish ages between those boundaries, linear interpolation was applied. The most important results of the stratigraphic work are published in Henrich et al. (1989), Sarnthein et al. (1992), Sarnthein and Altenbach (1995), and Weinelt et al. (1991). These publications also present ^{14}C ages, which were used wherever applicable. For all those cores that appear in this book but which have never been dated, ages were transferred from the next nearest core to the appropriate stratigraphic level.

In a further refinement, two absolute ages for the last glacial to interglacial transition were taken from the Greenland ice core record. The ages, 14,600 years bp and 11,600 ybp, are reported in Jouzel (1999), and were obtained by layer counting. The ages are calendar years before the present, and refer to the beginning of the first major warming after the Last Glacial Maximum (14,000 ybp), the Bølling Interstadial, and the second drastic warming after the younger Dryas time (11,600 ybp), which initiated the Holocene. These isotopic events are also seen in all Norwegian Sea box core delta ^{18}O records. At last, the newest 14C AMS age determinations from two long kastenlot cores with high resolution from the eastern Norwegian Sea were used to establish the ages of samples from cores 23071 and 23074 (Voelker, 1990).

14a Oxygen isotope stratigraphy is a concept that relies on the idea of isotopic fractionation in the hydrologic cycle to determine the ages of marine sediments. Water in the world's oceans does not contain only hydrogen with the atomic mass of 2 and oxygen with the atomic mass 16, but also small amounts of sister elements that deviate in their atomic mass but not in the number of electrons, so-called isotopes. One of them is oxygen-18. When water evaporates, the vapour is depleted in the heavier isotopes (including O-18) and relatively enriched in the lighter ones. Under present-day conditions, water that evaporates over the oceans and precipitates either as snow or rain over the continents, flows back towards the ocean. Therefore the ocean stays in isotopic equilibrium; the amount of heavy oxygen isotopes in it does not vary. But when the global ice sheets formed during the Pleistocene glaciations, the isotopically lighter evaporate (with less O-18 than the seawater) was taken out of the global ocean reservoir and stored on the continents as glacier ice. As a consequence, the remaining ocean water became relatively enriched in the heavier isotopes (O-18). Marine microorganisms, such as foraminifera that build calcareous shells out of seawater, record these isotope concentration changes in their skeletons and, with the help of instruments such as accelerator mass spectroscopes, these isotope ratios can be measured. Over the course of time, a vast set of marine isotope records from all over the world has been compiled, and all of those curves look somewhat similar in that they show a sawtooth-like shape. During the Pleistocene, oxygen isotope values have repeatedly changed from light (= similar to present-day conditions) to heavy (= fully glacial, very cold conditions), and then abruptly back to light during the deglaciations or terminations. The actual differences in the O-18 values are a result of the surface water temperature and the global ice volume. The ups-and-downs of isotope curves closely follow the changes in global insolation, which in turn are the result of the Earth's changing orbit around the sun, the so-called Milankovitch cycles. Through a process called orbital tuning, one can assign outstanding features of the O-18 curves to the maxima and minima of the global insolation curve. Since the ages of the latter can be accurately calculated to at least 300,000 years before the present, it is possible to use any oxygen isotope curve as a geochronometer.

14b The ages of cores KAL 23074 and 23071 are new, revised ages that in some cases differ dramatically from the stratigraphy known so far. The work was performed by Antje Voelker, who kindly provided the new chronology. Coarse fraction abundance data were also contributed by Elke Vogelsang. The cores have very high sedimentation rates and show the progression of the Weichselian Glaciation in maximum detail. The core chronology is based on oxygen isotope stratigraphy and numerous AMS (accelerator mass spectrometry) ^{14}C age measurements. This is the highest resolution record of oxygen isotope values and coarse fraction percentages from the Norwegian Sea, as far as it is known to this author. The complete work can be found in Voelker (1999).

15 As part of ongoing provenance studies in the Arctic Ocean, Dennis Darby and I have acquired more than 300 samples of terrestrial and shallow marine sediments from circum-Arctic sources which serve as 'fingerprints' of the local lithologies. The samples have been subjected to cluster analysis and form more than 40 groups of unique lithologic compositions. In samples between 30° West and 80° East north of 80 degrees, quartz grains outnumber feldspars by a wide margin. The maximum feldspar content among the minerals was in a sample

from north of Svalbard with 27%. But this is an exceptional sample. All others have much lower feldspar concentrations of much less than 7%. Therefore, any sample in the Norwegian Sea with at least 30% feldspar among the minerals is derived from Norway.

16 The samples were taken during a cruise of the American research vessel Northwind in 1963 and have these sample numbers: NW 63–89, NW 63–75, and NW 63–90. Pretty bold move of the Americans: exploring Soviet territory not two years after the Russian and American presidents brought the world to the brink of World War 3!

17 MacAyeal proposed that semi-rhythmical iceberg discharges could be the result of long-term gradual ice sheet growth and be triggered simply by a change in the ice sheet profile. Ice buildup will lead to a thickening of the ice sheet (in the case of the MacAyeal model, the Laurentide Ice Sheet over North America), and this will lead to a steeper height gradient from the thickest parts of the ice sheet towards its margin at sea level. The ice flow velocity towards the ocean would increase and large amounts of icebergs would calve off the ice front, dropping massive quantities of IRD along the way. After such a surge event, the ice sheet profile has thinned and needs time to regrow, before another iceberg surge can take place. During these regrowth phases, very small icebergs would calve off the ice sheet and, consequently, very low amounts of coarse IRD would be dropped in the ocean. These would be the fine-grained intervals between the IRD peaks recognised by Bond and colleagues in the North Atlantic. The important take-home point of the MacAyeal model is that it does not require a climate change (warming) as a triggering mechansim for iceberg surges. Rather, the cause of iceberg surges would be entirely physical, driven simply by the inescapable dynamic of a thickening ice sheet.

18 At the Third International Conference on Paleoceanography in Cambridge, UK, in 1989, I had the chance to sample two short piston cores with glaciomarine sediments from shallow depths off the island of Lewis, Outer Hebrides, for dropstone analysis. The samples (57-09-89, 326–331 cm core depth, and 57-09-33, 253–260 cm) contained abundant gneisses and granite fragments, few pieces of clastic sedimentary rocks and carbonates, and almost no quartzites and metapelites. The volcanic rocks were mostly grey and black basalts with a few dark green basalts. But sample 11 of core 23074 contains relatively high amounts of dark green basalts, some black types, and only one fragment of grey basalt which is most abundant in the Outer Hebrides. From their relatively fresh appearance, but differences in the colours, I conclude that the volcanics of sample 11 in core 23074 are of Tertiary age, but that they have an origin somewhere else than the northwestern British Isles. The next closest sources are volcanic rock outcrops on the Faeroer Islands or on the bottom of the southern Norwegian shelf (Bygge et al., 1980). The ice drift direction at this time (25.6–25 ka) was towards the north.

19 The lithology of basalts from Franz Josef Land was studied in a shallow marine sediment sample from the vicinity of Franz Josef Land. The sample, NP94 31-GC-1, was kindly provided by Anders Solheim of the Norwegian Polar Institute.

It contained numerous fragments of a greenish-grey basalt. It has a fine crystalline (10–50 μm size crystals) groundmass with large phenocrysts of plagioclase. The mineral composition is green and yellow – clear clinopyroxene, clear plagioclase and black opaques. The basalt looks fresh and contains pores which are partly filled with diffuse green matter. There are also rests of black obsidian (volcanic glass). This rock does not occur in any of the Norwegian Greenland Sea IRD samples, and thus debris from Franz Josef Land left no recognisable trace in the marine samples.

20 The decay of the western Barents Ice Sheet took place in two phases: first, the southwestern part of the ice sheet disintegrated. This happened shortly after the last glacial maximum, based on the oldest sample from west of the Barents shelf, 23261, 189–190 cm, with an age of 17450–17360 calendar years bp. This first pulse produced the very high clastic rock percentages in >500 μm IRD at the Vøring-Plateau, and the clastics were dominated by grey siltstones. After much of the ice was gone, disintegration continued further north, and the icebergs rafted preferably black shales into the Greenland Sea. This second phase, which lasted at least until 13,100 cal. years bp (57% clastic rock IRD in core 23258, 588–591 cm), is not recognised in the Norwegian Sea, and thus the icebergs must have taken a more westerly drift route through the Greenland and Iceland seas.

Interestingly, the thickness of the first iceberg pulse from the southwestern Barents shelf is not much different between the Barents slope cores (10–20 cm thickness) and those in the Norwegian Sea (less than 10 cm). This points to a fast southward drift in relatively cold waters, extending particle drop-out until the Vøring-Plateau. To a similar conclusion came Sarnthein et al. (1992), who noticed a light meltwater signal in their O-18 records in the northern Norwegian Sea. But this meltwater signal (time slice 13.6 ka, ^{14}C age) was not directly at the Barents Ice Sheet margin, but occurred 100–200 km to the west. They attributed this to the melting of icebergs at some distance as the result of an encounter with warmer waters that flowed northward through the central Norwegian Greenland Sea. Both observations can be folded into this paleoceanographic model: The southward drift of icebergs from the Barents Ice Sheet was stopped by the physical force of the northbound currents which, by means of their higher temperature also caused increasing melting of the icebergs.

21 The formula is: the depth of the global ocean today (3800 m) times its salinity (36.6 per mil) equals the depth of the ice age global ocean ($3800 - 130$ m $=$ 3670 m) times its unknown salinity: 3800 m $\times$ $36.6 = 3670$ m $\times$ $S_{ice\ age}$. $S_{ice\ age}$ is 37.9 per mil.

22 Dierk Hebbeln and Gerold Wefer placed sediment traps in Fram Strait. The traps were positioned so that they would catch sediments deposited under three different climatic situations: (1) under permanently ice covered and cold waters, the East Greenland Current, (2) under a seasonally variable ice cover in cold waters, and (3) under seasonally variable ice cover in temperate waters. The use of the term 'temperate' refers to the waters in the West Spitsbergen Current which have a temperature of 2–4°C. These temperatures are certainly not inviting for swimmers, but they have an enormous effect on the sedimentation

rates. Ice drifting in cold waters releases very little sediment. In fact, the accumulation rates under the East Greenland Current are the same low rates that dominate in the Arctic Ocean, about 1 mm/kyr. But the trap underneath the only slightly warmer West Spitsbergen Current received so much sediment that one of the revolving glass cylinders spilled over. The rates of sedimentation in this scenario can easily approach 1 m/kyr or more. Therefore, even a subtle increase of the surface water temperature to a few degrees above zero can result in an increase of the rate of ice rafted sedimentation of several orders of magnitude.

23 The **vitrinite reflectance ratio** is a numerical value that characterises different types of coal. For most commercially-used coal deposits, these values are precisely known. We (Joachim Koch, Robert Spielhagen, Michaela Kubisch, Jörn Thiede and I), performed a study to find the sources of coal fragments that we observed in our glacimarine sediments from the eastern Arctic Ocean (Kubisch), the Fram Strait (Spielhagen), and the Norwegian Sea. Joachim Koch of the Bundesanstalt für Geowissenschaften und Rohstoffe in Hannover, Germany, measured the reflectance ratios and concluded that the coals were not from Svalbard, but must have a different source further to the east, possibly in Franz Josef Land or northern Siberia. This is further corroborated by highest coal concentrations in the eastern Arctic Ocean, gradually declining along a path through Fram Strait into the Norwegian Sea. The work was published in a book edited by Julian Dowdeswell and James Scourse, *Glaciomarine Environments: Processes and Sediments*, as Bischof et al. in 1990.

24 The shallow marine samples are ice rafted from sources close to the sample sites. Comparison of the petrographic content of these samples with the published literature and maps of nearby geologic formations show that the debris in these samples is from within a radius of 200 km. Other sample types are glacial tills, both terrestrial and marine, raised marine sediments, river sands, proglacial delta deposits and loose surface sediment samples. They were kindly provided by John England, Donald Lemmen, Robert Gilbert, Weston Blake Jr., Peta Mudie, Iris Hardy, Jean-Serge Vincent, Svend Funder, Carl-Egede Bøggild, Anders Solheim, Robert Spielhagen, Scott Lamoureux, Arthur Dyke, Douglas Hodgson and Bobbi Conard.

25 Our approach to tracing IRD back to its sources involves two steps: (1) the characterisation of potential source areas, and (2) matching the sediment core samples to the source areas with multivariate statistics. The details of these steps can be found in Darby and Bischof (1996). The first step consists of characterising potential source areas for Fe oxide grain composition types and assemblages of coarse (over 0.25 mm) IRD grain types. Then the coarse IRD grains are matched to a source by discriminant function analysis (DFA). The coarse grains provide ground-truth, or confirmation, of the results from the iron oxide grains: of all land areas bordering the Arctic Ocean, carbonates are specifically concentrated in the Canadian Arctic Archipelago. Victoria Island, for example, consists of more than 80% of its surface of carbonates. On the other hand, carbonate exposures are very limited in the Eurasian Arctic where

clastic sedimentary or crystalline rocks dominate. This is clearly reflected in our source area samples: till samples from Victoria Island contain the highest carbonate percentages of up to 52% (in percent of lithogenic grains over 250 microns), whereas the Eurasian Arctic samples, which are shallow marine sediments produced by local ice rafting, contain only 2–5% carbonates. Therefore, elevated carbonate percentages in an Arctic Ocean IRD sample clearly identify the North American Arctic, and more specifically the south-western part of the Canadian Arctic islands, as the major IRD source. This confirmation is needed to put the results from the Fe oxide grains in the right perspective. The procedure involves two separate data sets, the grain type counts and the microprobe data on Fe oxide grains. The grain types (lithic and mineral fragments) in the over 250 micron fraction are microscopically identified and categorised into one of 72 types. More than 500 grains per sample are counted. The Fe oxide grains are separated by hand magnet and a Frantz magnetic separator from the 45–250 micron fraction and mounted in a transparent resin, polished, and identified by reflected light microscopy. Each grain is numbered on a photo and analysed by microprobe for 12 elements using a beam diameter about half that of the smallest grain to acquire an average grain composition. Grains larger than three times the beam diameter (approximately 20 microns) are analysed in more than one spot and the results averaged. The source areas are determined by unique grain type composition and constrained by geologic provinces or areas of present or former ice masses. Unique grain types are determined by cluster analysis and refined by DFA using a probability of 0.95 for group membership. Once unique source areas are determined, the Fe oxide grains for each of nine mineral types are clustered and tested by DFA in order to obtain composition types with very limited variance for each Fe oxide mineral (Darby and Bischof, 1996). The core samples are compared to source areas for the two data sets separately. DFA can classify core samples into a source area group and provide an indication of those sources that are most similar (see Darby and Bischof, 1996 for a more complete discussion). Definitive sources and proportions of multiple sources in a core sample can only be obtained by matching both the coarse sand fraction and individual, single Fe oxide grains to their sources by DFA.

26 Icebergs and sea ice drifting through the western Fram Strait along the East Greenland continental margin during the past 30,000 years originated from rapidly-changing sources in northern Eurasia between 10 and 110° E and the Canadian Arctic Archipelago. The sources were determined by discriminant function analysis (DFA) fingerprinting, a statistical technique that compares the composition of coarse ice rafted debris (IRD) in three ^{14}C AMS dated box cores from the western Arctic Ocean (92 BC 17), the central Arctic Ocean (94 BC 17), and Fram Strait (PS 1230-1) to the composition of slightly over 300 source area samples from the circum-Arctic. These source area samples are mostly glaciomarine sediment which itself consists of lithologies that were transported some distance, but which strongly represents local rock types. From comparison of the samples' composition with nearby bedrock, the source sample granules

were found to be from a 200 km radius around the sample site, and thus are representative of a relatively wide range of rock types available for ice rafting in the Arctic Ocean.

DFA of the >250 μm lithic grains in core PS 1230 determined that 21 samples came from northern Eurasia, 10 from North America, and three from northeast Greenland, while 15 were from unknown sources. The presence of a unique multicoloured phyllite with randomly oriented amphibole crystals from Yelverton Inlet, northwest Ellesmere Island,[26a] in 18 samples illustrates the mixed nature of the IRD and proves that most samples were deposited by ice from both the Eurasian and the North American sides of the Arctic Ocean. Reinspection of the DFA matches showed that the discrete percent values of some IRD lithologies fall outside the data range of the assigned source group. In some instances, these differences can add up to 30%, which clearly demonstrates that the samples are always mixtures from several sources with the determined source dominating. The percentage of debris from the assigned source typically varies from 60–90%. The Fe oxide grain matches confirm this with multiple sources from different parts of the Arctic Ocean's periphery contributing at any time. The closest, and in some instances even perfect, matches between sample and source area data were obtained for core 92 BC 17 from the Northwind Ridge. The IRD sources at this location are exclusively located in the southern Queen Elizabeth Islands, Banks and Victoria islands, and in the northern District of Mackenzie. Thus, icebergs from the northwestern Laurentide ice front left a clearly recognisable footprint in the Northwind Ridge, and the extent of mixing with IRD from non-Laurentide sources has been minimal. The most notable feature of the IRD record is the carbonate peak between 15 and 5ka with a maximum value of 50% at 10ka (Fig. 5.11). The same carbonate peak is also seen in the two other cores, albeit at progressively lower maximum values. The gradual decline of the carbonate peak height from 50% in 92BC17 to 34% in 94BC17 near the North Pole, and finally to less than 11% at PS 1230, Fram Strait, is caused by increasing numbers of icebergs from the combined Barents/Kara, Innuitian and Greenland ice sheets. The differences of the IRD lithologies between the three areas are illustrated in Fig. 5.12 which shows quartz, carbonates and cherts, clastic sedimentary rocks (including coal fragments), and crystalline rocks plotted as ratios. In the Northwind Ridge area (black squares), the ratios of clastics to crystallines are highly variable and the ratios of quartz to carbonates and cherts are less than 4 with a concentration of data points around 1. Near the North Pole (blue squares), the ratios of clastics to crystallines are generally less than 1 but the ratios of quartz to carbonates and cherts are highly variable, ranging from 1 to 15. Data points from Fram Strait (pink dots) do not overlap with those from the Northwind Ridge. Crystalline rocks and quartz are becoming increasingly abundant along our transect across the Arctic Ocean.

The phyllite from Yelverton Inlet, northwest Ellesmere Island, was thus far only found in two long piston cores from the southern Alpha Ridge (Bischof et al., 1996; Bischof and Darby, 1997), where it is associated with abundant mafic crystalline rocks. This phyllite and the associated mafics serve as witness that ice from northern Ellesmere Island drifted out of the Arctic on a curved path towards Fram Strait and did not circulate clockwise as would be dictated by the concept of the Beaufort Gyre. The Yelverton Inlet phyllite is particularly frequent during the lower part of the Fram Strait core between 30 and 21 ka and could reflect the growth phase of the Innuitian Ice Sheet. During the following two kyr the debris came from northeast Greenland and Ellef Ringnes Island, and until 15.8 ka it originated from Ellef Ringnes Island, Svalbard, Taymyr Peninsula and the New Siberian Islands. From 15.8 until 12.8 ka, the IRD originated primarily from Svalbard, and until 8.3 ka it came primarily from Axel Heiberg Island. The sources of Holocene IRD are mostly in northern Eurasia.

The >250 μm record of PS 1230-1 contains no evidence for ice from the north slope of Alaska, the Chukchi Sea or the East Siberian shelves. It is possible that a clastic sedimentary rock peak beginning at 15.5 ka contains evidence for the receding Barents Ice Sheet, and that a subsequent carbonate peak at 10.5 ka contains the remnants of the Laurentide Ice Sheet.

26a This peculiar rock was first seen in a long piston core from north of Ellesmere Island (core FL-561, results published in Bischof et al., 1996). The rock occurs in small quantities practically throughout the core in conjunction with large quantities of multicoloured crystalline rocks. The phyllite is special because it has relatively large amphibole blades stuck at random orientation in an otherwise plane-parallel foliated micaceous matrix. The rock is truly unique and was described in an unpublished technical report by Bjørnerud and Ohta. The phyllite occurs in a wide range of different colours, including grey, silver, shades of green, red, purple and even blue. It is a very eye-catching rock, difficult to miss in microscopic examination. However, apart from core FL-561 and another core from approximately the same location (FL-542, results published in Bischof and Darby, 1997), this rock was not seen once in all of the other cores from the central and western Arctic Ocean – and that despite the fact that I routinely scanned across the entire sample after the grain counts on a sample split were completed. This ensures that this particular phyllite was not simply overlooked, but really was not transported by icebergs or sea ice to the other Arctic core sites. Therefore, the distribution of this rock alone allows us to state that the main direction of ice drift north of Ellesmere Island was, and probably still is, to the east, towards Fram Strait.

27 The grain frequencies were obtained by counting grains in a calculated sample split, and multiplying the number by the split factor. The counts were performed by a student, Christina Majer. Box core 92BC6 was selected for this method because it offered large volume samples. Whenever smaller sample volumes are analysed this way, the grains may be from a coarse patch or a mud clast. But samples that were cut across the entire width of a box core avoid this problem and give an integrated and more realistic value of coarseness.

References

Agnew, T.A. and Le, H. (1996). A more comprehensive view of the dynamics of the Arctic sea ice as revealed through 85.5 GH_2 SSM/I imagery. *NSIDC Notes*; **19**.

Anderberg, M.R. (1973). *Cluster Analysis for Applications*. New York: Academic Press Inc.

Andersen, B.G. (1990). Cenozoic glacier fluctuations in polar regions – Terrestrial records from Antarctica and the North Atlantic sector of the Arctic. In: U. Bleil and J. Thiede (Eds) *Geological History of the Polar Oceans: Arctic versus Antarctic*. Amsterdam: Kluwer Academic Press; 245–254.

Anderson, John B. (1983). Ancient glacial-marine deposits: Their spatial and temporal distribution. In: Bruce F. Molnia (Ed.) *Glacial–Marine Sedimentation*. New York: Plenum; 3–94.

Anderson, John B., Brake, C., Domack, E., Myers, N. and Wright, R. (1983). Development of a polar glacial–marine sedimentation model from Antarctic Quaternary deposits and glaciological information. In: Bruce F. Molnia (Ed.) *Glacial–Marine Sedimentation*. New York: Plenum; 233–264.

Andrews, J.T., Erlenkeuser, H., Evans, L.W., Briggs, W.M. and Jull, J.T. (1991). Meltwater and deglaciation, SE Baffin Shelf (NE Margin Laurentide Ice Sheet). between 13.5 and 7 ka: from O and C stable isotope data. *Paleoceanography*; **6(5)**: 621–637.

Baumann, K.H., Lackschewitz, K.S., Mangerud, J., Spielhagen, R., Wolf-Welling, T.C.W., Henrich, R. and Kassens, H. (1995). Reflection of Scandinavian ice sheet fluctuations in Norwegian Sea sediments during the past 150,000 years. *Quat. Res.*; **43**: 185–197.

Bell, Robert, 1886. Observations on the geology, zoology, and botany of Hudson's Strait and Bay made in 1885. *Geological Survey of Canada Annual Report* (new series)., **vol. 1**, pages 1DD-27DD (describes wind-blown dust on the ice).

Berggren, W.A., Kent, D.V., Swisher, C.C. III, Aubry, M.-P. (1995). A revised Cenozoic geochronology and chronostratigraphy. In: W.A. Berggren, et al. (Eds) *Geochronology, Time Scales and Global Stratigraphic Correlation*. SEPM Special Publication, 54: 129–212.

Bischof, J.F. (1994). The decay of the Barents ice sheet as documented in Nordic seas' ice-rafted debris. *Marine Geology*; **117**: 35–55.

Bischof, J.F. (1990). Dropstones in the Norwegian-Greenland Sea – Indications of Late Quaternary circulation patterns? In: U. Bleil and J. Thiede (Eds) *Geological History of the Polar Oceans: Arctic versus Antarctic*. NATO ASI Series C: Mathematical and Physical Sciences, Vol. **308**. Dortrecht: Kluwer Academic; 499–518.

Bischof, J.F., Clark, D.L. and Vincent, J.-S. (1996). Origin of ice-rafted debris: Pleistocene paleoceanography in the western Arctic Ocean. *Paleoceanography*, **11(6)**: 743–756.

Bischof, J.F. and Darby, D.A. (1999). Quaternary ice transport in the Canadian Arctic and extent of Late Wisconsin Glaciation in the Queen Elizabeth Islands. *Can. J. of Earth Sciences*, 1999; **36**: 2007–2022.

Bischof, Jens F., Darby, D.A. and Majer, C. (1999). Very high resolution record of Late Pleistocene to Holocene ice rafting and glacio-fluvial meltwater discharge from the northern Chukchi Sea. Supplement to EOS, *Transactions*, AGU, **80(46)**: 16 November 1999. Poster OS31B-08, p. F534.

Bischof, J.F., Koch, J., Kubisch, M., Spielhagen, R.F. and Thiede, J. (1990). Nordic Seas surface ice drift reconstructions: Evidence from ice rafted coal fragments during oxygen isotope stage 6. In: J.A. Dowdeswell, and J.D. Scourse (Eds) *Glaciomarine Environments: Processes and Sediments*. Geological Society Special Publication No. **53**: 235–251.

Bischof, Jens F., Lund, Jens-Jörgen, Ecke, Hans-Herrman (1997). Palynomorphs of ice rafted clastic sedimentary rocks in Late Quaternary glacial marine sediments of the Norwegian Sea as provenance indicators. *Palaeogeography, Palaeoclimatology, Palaeoecology*, **vol. 129**, No.3–4, May 23, 1997: 329–360.

Bleil, U. and Thiede, J. (Eds) (1990). *Geological History of the Polar Oceans: Arctic versus Antarctic*. NATO ASI Series C: Mathematical and Physical Sciences, Vol. **308**. Dortrecht: Kluwer Academic Publishers.

Boese, Margot (1990). Reconstruction of ice flow directions south of the Baltic Sea during the Saalian and Weichselian glaciations. *Boreas*, **19**: 217–226.

Bond, G.C. and Lotti, R. (1995). Iceberg discharges into the North Atlantic on millenial time scales during the last glaciation. *Science*, **267**: 1005–1009.

Bond, G., Broecker, W., Johnsen, S., McManus, J., Labeyrie, L., Jouzel, J. and Bonani, G. (1993). Correlations between climate records from North Atlantic sediments and Greenland Ice. *Nature*, **365**: 143–147.

Boulton, G.S., Smith, G.D., Jones, A.S. and Newsome, J. (1985). Glacial geology and glaciology of the last mid-latitude ice sheets. *J. Geol. Soc. London*, **142**: 447–474 (very comprehensive, several models of late Quaternary ice sheets, one coalescent between Great Britain and Scandinavia, and one open).

Brown, Robert (1875). On the physical structure of Greenland. In: Royal Geographic Society, *Arctic papers for the expedition in 1875*. London: John Murray; 1–74. (description of rock avalanches onto sea ice).

Bygge,T., Prestvik, T., Rokoengen, K. (1980). Lower Tertiary volcanic rocks off Kristiansund, Mid-Norway. *Marine Geology*, **35**, 277–286.

Carmack, Eddy C., Aagaard, K., Swift, James H., Macdonald, Robie W., McLaughlin, E. Fiona A., Jones, Peter., Perkin, Ronald G., Smith, John N., Ellis, Katherine M. and Killius, Linus R. (1997). Changes in temperature and tracer distributions within the Arctic Ocean: Results from the 1994 Arctic Ocean section. In Patricia A. Wheeler (guest editor)., *Topical Studies in Oceanography*. 1994 Arctic Ocean Section. Deep Sea Research, part II, volume 44, No. 8: 1487–1502.

Cavalieri, D.J., Gloersen, P., Parkinson, C.L., Comiso, J.C. and Zwally, H.J. (1997). Observed hemispheric asymmetry in global sea ice changes. *Science*, **278**, 1104–1106.

Churkin, M. Jr., Soleimani, G., Carter, C. and Robinson, R. (1981). Geology of the Soviet Arctic: Kola Peninsula to Lena River, Chapter V, Franz Josef Land – Area IV. In: A.E.M. Nairn, M. Churkin, Jr. and F.G. Stehli (Eds) *The Ocean Basins and Margins, Volume 5, The Arctic Ocean*. New York and London: Plenum; 348–349.

Clark, D.L., Whitman, R.R., Morgan, K.A. and Mackey, S.D. (1980). Stratigraphy and glacial–marine sediments of the Amerasian Basin, Central Arctic Ocean. *Geological Society of America Special Paper* **181**; Boulder, Co.

Clark, D.L. 1990. Arctic Ocean ice cover; Geologic history and climatic significance. In: A. Grantz, L. Johnson and J.F. Sweeney (Eds) *The Arctic Ocean region*. Boulder, CO: Geological Society of America, Boulder, Colorado, 53–62.

Clark, Peter U., Alley, R.B. and Pollard, D. (1999). Northern hemisphere ice-sheet influences on global climate change. *Science*, **286**: 1104–1111 (new ice sheet thickness reconstruction for northern hemisphere, plus several good isotope ice volume records).

Colony, R. and Thorndike, A.S. (1985). Sea ice motion as a drunkard's walk. *Journal of Geophysical Research*, **90(C1)**: 965–974.

Cullen, H., deMenocal, P., Hemming, S., Hemming, G., and Brown, F. (1997). Possible role of climate in the collapse of the Akkadian Empire: Evidence from Gulf of Oman marine sediments. *Supplement to EOS Transactions*, **78(46)**: abstract and poster OS41D-9, F376.

Dansgaard, W., Johnsen, S.J., Clausen, H.B., Dahl-Jensen, D., Gundestrup, N.S., Hammer, C.U., Hvidberg, C.S., Steffensen, J.P., Sveinbjörnsdottir, A.E., Jouzel, J. and Bond, G. (1993). Evidence for general instability of past climate from a 250-kyr ice-core record. *Nature*, **364**: 218–220.

Darby, D.A., Bischof, J.F., Cutter, G.A., deVernal, A., Hillaire-Marcel, C., de Bilodeau, G., McManus, J., Dwyer, G.S., Polyak, L., Poore, R.Z. and Ostermann, L. (principal investigators). (2000). *Collaborative Research Project: Paleoceanographic Records of Western Arctic Shelf-Basin Interactions*. Ongoing Research Project in the SBI (Shelf-Basin-Interactions). Initiative of the Arctic System Science division of the Office of Polar Programs, National Science Foundation.

Darby, D.A., Bischof, J.F. and Jones, G.A. (1997). Radiocarbon chronology of depositional regimes in the western Arctic Ocean. *Deep Sea Research II*, **44(8)**: 1745–1757.

Darby, D.A. and Bischof, J.F. 1996. A stastistical approach to source determination of lithic and Fe-oxide grains: An example from the Alpha Ridge, Arctic Ocean. *Journal of Sedimentary Research*, **66(3)**: 599–607.

Davis, John C., 1986. *Statistics and Data Analysis in Geology*. New York: John Wiley & Sons.

Dowdeswell, J.A. (1995). Deep Pleistocene iceberg plowmarks on the Yermak Plateau: Sidescan and 3.5 rhz evidence for thick calving ice fronts and a possible marine ice sheet in the Arctic Ocean. Comment. *Geology*: 476–477.

Dowdeswell, J.A. and Dowdeswell, E.K. (1989). Debris in icebergs and rates of glacimarine sedimentation: Observations from Spitsbergen and a simple model. *Journal of Geology*, **97**: 221–231.

Dowdeswell, J.A., Maslin, M.A., Andrews, J.T. and McCave, I.N. (1995). Iceberg production, debris rafting, and the extent and thickness of Henrich layers (H-1, H-2). in North Atlantic sediments. *Geology*; **23(4)**: 304–304.

Dowdeswell, Julian A. and James D. Scourse (Eds) (1990). *Glacimarine Environments: Processes and Sediments*. Geological Society Special Publication No. 53, London.

Dowedeswell, J.A., Villinger, H., Whittington, R.J. and Marienfeld, P. (1993). Iceberg scouring in Scoresby Fund and on the East Greenland continental shelf. *Marine Geology*; **111**: 37–53.

Dowdeswell, J.A., Whittington, R.J. and Hodgkins, R. (1992). The sizes, frequencies, and freeboards of East Greenland icebergs observed using ship radar and sextant. *Journal of Geophysical Research*, **97**: 3515–3528.

Drewry, D.J. 1986. *Glacial Geologic Processes*. London: Arnold.

Dyke, A.S. (1998). Holocene delevelling of Devon Island, Arctic Canada: Implications for ice sheet geometry and crustal response. *Canadian Journal of Earth Sciences*, **35(8)**: 885–904 (postglacial uplift patterns indicate ice centre to the NW of Devon Island, therefore the existence of an Innuitian Ice Sheet).

Dyke, A.S. (1999). Last glacial maximum and deglaciation of Devon Island, Arctic Canada: Support for an Innuitian Ice Sheet. *Quaternary Science Reviews*, **18**: 393–420 (extensive review of all available evidence for and against an Innuitian Ice Sheet. However, the author shies away from actually sketching the ice sheet. We did in the most recent issue of *CJES*).

Eicken, H., Kolatschek, J., Freitag, J., Lindemann, F., Kassens, H. and Dmitrenko, I. (2000). A key source area and constraints on entrainment for basin-scale sediment transport by Arctic sea ice. *Geophys. Res Lett.*; **27**: 1919.

Eicken, H., Viehoff, T., Martin, T.J., Kolatschek, J., Alexandrow, V., and Reimnitz, E. (1994). Studies of clean and sediment-laden ice in the Laptev Sea. In: D.K. Fütterer (Ed.) *The Expedition ARCTIC '93 Leg ARK-IX/4 of RV 'POLARSTERN'*. Berichte zur Polarforschung – Reports on Polar Research, **149**, Alfred Wegener Institute for Polar and Marine Research: 62–70.

Elverhøi, A., Antonsen, P., Flood, S.B., Solheim, A. and Vullstad, A.A. (1988). Shallow bedrock geology. In: *The physical environment; Western Barents Sea, 1:1.500.000*. Norsk Pol. Inst. Skrifter **179 D**.

Elverhøi, A., Fjeldskaar, W., Solheim, A., Nyland-Berg, M., and Russwurm, L., (1993). The Barents Sea ice sheet – a model of ice growth and decay during the last glacial maximum. *Quaternary Science Reviews*, **12**: 863–873.

Elverhøi, A., Nyland-Berg, M., Russwurm, L. and Solheim, A. (1990). Late Weichselian ice recession in the Central Barents Sea. In: U. Bleil and J. Thiede

(Eds) *Geological History of the Polar Oceans: Arctic vs. Antarctic*. Dordrecht: Kluwer Academic Publishers; 289–307.

Eyles, C.H., Eyles, N. and Lagoe, M.B. (1991). The Yakataga Formation: A late Miocene to Pleistocene record of temperate glacial marine sedimentation in the Gulf of Alaska. *Geological Society of America*, Special Paper 261 (exposed thick unit of ice rafted sediments in a temperate setting, 5 km thickness, glacial marine sedimentation began in the late Miocene).

Forman, S.L., Lubinski, D., Miller, G.H., Snyder, J., Matishov, G., Korsun, S., and Myslivets, V. (1995). Postglacial emergence and distribution of late Weichselian ice-sheet loads in the northern Barents and Kara Seas, Russia. *Geology*, **23(2)**: 113–116.

Fulton, R.J. (1989). Foreword to the Quaternary Geology of Canada and Greenland. In: R.J. Fulton (Ed.) *Quaternary Geology of Canada and Greenland*. Geological Survey of Canada, Geology of Canada, no. **1**: p. 1–11. (Also in: *Geological Survey of America, The Geology of North America*, Vol. **K-1**, 1–12).

Fütterer, D.K., Grobe, H. and Grünig, S. (1988). Quaternary sediment patterns in the Weddell Sea: Relations and environmental conditions. *Paleoceanography*, **3**: 551–561 (icebergs drift mainly with the Antarctic Coastal Current, as evidenced by high IRD content there).

Geiger, C.A., Ackley, S.A. and Hibler, W.D. III (1998). Sea ice drift and deformation processes in the western Weddell Sea. In: Martin O. Jeffries (Ed). *Antarctic Sea Ice: Physical Processes, Interactions and Variability*. Washington: Antarctic Research Series, American Geophysical Union; 141–160.

Gerlach, S.A., Thiede, J., Graf, G. and Werner, F. (1986). Forschungsschiffahrt Meteor Reise 2 vom 19 Juni bis 16 Juli 1986. *Ber. Sonderforschungsbereich Univ. Kiel*, **313(4)**: 35–81.

Gilbert, R. (1983). Sedimentary processes of Canadian Arctic Fjords. *Sedimentary Geology*, **36**: 147–175.

Gilbert, R. (1990). Rafting in glaciomarine environments. In: J.A. Dowdeswell and J.D. Scourse (Eds) *Glacimarine Environments: Processes and Sediments*. London: Geological Society Special Publication No. 53; 105–120.

Grousset, F.E., Pujol, C., Labeyrie, L., Auffret, G. and Boelart, A. (2000). Were the North Atlantic Heinrich events triggered by the behavior of the European ice sheets? *Geology*, **28(2)**: 123–126 (European ice sheets disintegrate before the Laurentide, again in sequence).

Häkkinen, S. (1993). An arctic source for the Great Salinity Anomaly: A simulation of the arctic ice-ocean system for 1955–1975. *J. Geophys. Res.*, **98**: 16,397–16,410.

Hancock, J.M. (1984). Cretaceous. In: K.W. Glennie (Ed.) *Introduction to the Petroleum Geology of the North Sea*. Oxford: Blackwell Scientific Publications; 133–150.

Harder, M., Lernke, P. and Hilmer, M. (1998). Simulation of sea ice transport through Fram Strait: Natural variability and sensitivity to forcing. *J. Geophys. Res. (Oceans)*.; **103**: 5595–5606.

Hebbeln, Dierk and Wefer, Gerold, 1991. Effects of ice coverage and ice-rafted material on sedimentation in the Fram Strait. Nature, Vol. 350, no. 6317, 4th April, 1991: 409–411.

Heinrich, H. (1988). Origin and consequences of cyclic ice rafting in the Northeast Atlantic Ocean during the past 130,000 years. *Quaternary Research*, **29**: 142–152.

Henrich, R. (1990). Cycles, rhythms, and events in Quaternary Arctic and Antarctic glaciomarine deposits. In: U. Bleil and J. Thiede (Eds) *The Geological History of Cenozoic Polar Oceans: Arctic versus Antarctic*. NATO ASI Series C: Mathematical and Physical Sciences – Vol. **308**. Dortrecht: Kluver Academic; 213–244.

Henrich, R., Kassens, H., Vogelsang, E. and Thiede, J. 1989. Sedimentary facies of glacial – interglacial cycles in the Norwegian Sea during the last 350 ka. *Marine Geology*, **86**: 283–319.

Hibler, W.D. (1979). A dynamic thermodynamic sea ice model. *J. Phys. Oceanogr.*; **9**: 815–846.

Hilmer, M., Harder, M. and Lemke, P. (1998). Sea ice transport: A highly variable link between Arctic and North Atlantic. *Geophys. Res. Lett.*; **25**: 3359–3362.

Hjelle, A. and Ohta, Y. (1974). Contribution to the geology of northwestern Spitsbergen. *Norsk Polar Inst. Skrifter*, **158**.

Hodgson, D.A. (1989). Quaternary geology of the Queen Elizabeth Islands, In R.J. Fulton (Ed.) *Quaternary Geology of Canada and Greenland*. Geological Survey of Canada, Geology of Canada, no. **1**: p. 441–478. (Also in: Geological Survey of America, The Geology of North America, Vol. **K-1**).

Hult, J.L. and Ostrander, N.C. (1973). *Antarctic Icebergs as a Global Fresh Water Resource*. Technical Report R-1255-NSF, prepared for the National Science Foundation.

Jakobsson, M., Løvlie, R. and Backman, J. (1998). Late Pleistocene ice grounding on the Lomonosov Ridge, Central Arctic Ocean? *Supplement to EOS Transactions*, AGU volume 79, number 45: Poster OS12C-24, F438.

Jansen, E. 1992. Following iceberg footprints. *Nature*, Vol. **360**: 212–213.

Jensen, H., Loset, S. and Lovas, M. (1991). Sampling of iceberg ice. In G. Krause, J. Meincke and H.J. Schwartz (Eds) *Scientific Cruise Reports of Arctic Expeditions ARK VI/1-4 of RV 'Polarstern' in 1989*, Reports on Polar Research, no. 87. Bremerhaven: Alfred Wegener Institute for Polar and Marine Research, Bremerhaven, 10–11.

Johannessen, O.M., Miles, M. and Bjorgo, E. (1995). The Arctic's shrinking sea ice, *Nature*, **376**: 126–127.

Johannessen, Ola M., Sandven, S., Budgell, W.P., Johannessen, J.A. and Shuchman, R.A. (1994). Observation and simulation of ice tongues and vortex pairs in the marginal ice zone. In: O.M. Johannessen, R.D. Muench, and J.E. Overland, (Eds) *The Polar Oceans and Their Role in Shaping the Global Environment*. The Nansen Centennial Volume. American Geophysical Union, Geophysical Monograph 85: 109–136.

Johannessen, O.M., Muench, R.D. and Overland, J.E. (Eds) (1994). *The Polar Oceans and Their Role in Shaping the Global Environment*. The Nansen Centennial Volume. American Geophysical Union, Geophysical Monograph 85.

Johannessen, O.M. (1986). *Brief overview of the physical oceanography*. In: B.G. Hurdle (Ed.) *The Nordic Seas*. New York: Springer Verlag 103–127.

Johnson, M.A., Proshutinsky, A.Y. and Polyakov, I.V. (1999). Atmospheric patterns forcing two regimes of Arctic circulation: A return to anticyclonic conditions? *Geophysical Research Letters*, **26(11)**: 1621–1624.

Jouzel, J. (1999). Calibrating the isotopic Paleothermometer. *Science*, **286**, 29 October 1999; *Perspectives: Geosciences*: 910–911.

Kassens, H., Piepenburg, D. Thiede, J. Timokhov, L. Hubberten, H.-W. and Priamikov, S.M. (Eds) (1995). *Russian – German Cooperation: Laptev Sea System.* Berichte zur Polarforschung, Reports on Polar Research, 176. To be obtained from The Alfred Wegener Institute for Polar and Marine Research, Columbusstrasse, D-27568 Bremerhaven, Germany.

Krause, G., Meincke, J. and Schwartz, H.J. (Eds) (1991). *Reports on Polar Research, no. 87.* Scientific Cruise reports of Arctic Expeditions ARK VI/1-4 of RV 'Polarstern' in 1989. Bremerhaven: Alfred Wegener Institute for Polar and Marine Research.

Kreyscher, M., Harder, M. and Lemke, P. (1997). First results of the Sea Ice Model Intercomparison Project (SIMIP). *Ann.Glaciol.*; **25**: 8–11.

Kreyscher, M., Harder, M., Lemke, P. and Flato, G.M. (2000). Results of the Sea Ice Model Intercomparison Project: Evaluation of sea ice rheology schemes for use in climate simulations. *J. Geophys. Res.*; in press.

Krissek, L.A. (1989). Late Cenozoic records of ice-rafting at ODP sites 642, 643 and 644, Norwegian Sea: Onset, chronology and characteristics of glacial/interglacial fluctuations. In: *Proceedings of the Ocean Drilling Program, Scientific Results, Volume 104*: 61–74.

Kronborg C., Bender H., Bjerre R., Friborg R., Jacobsen H.O., Kristiansen L., Rasmussen P., Soerensen P.R. and Larsen G. (1990). Glacial stratigraphy of East and Central Jutland. *Boreas*, **19**: 273–287 (erratic descriptions and counts, mostly flint, igneous, some sedimentary rocks, and quartz, two major sources, one in the Scandinavian Shield, Precambrian, and another one in the Baltic depression with Precambrian to Paleozoic sediments, chalk, and flint).

Kudrass, H.R. (1973). Sedimentation am Kontinentalhang vor Portugal und Marokko im Spätpleistozän und Holozän. *Meteor Forschungsergebnisse*, **13**: 1–63.

Kwok, R., and Rothrock, D. (1999). Variability of Fram Strait ice flux and North Atlantic Oscillation. *Journal of Geophysical Research C: Oceans*, **104(3)**: 5177–5189 (satellite based calculations of the annual average ice flux, by volume and area, high fluctuations around the long term, 18 year mean correlate with sea level atmospheric pressure).

Lamy, F., Hebbeln, D., Röhl, U. and Wefer, G. (1999). Holocene climate variability in southern South America: A marine record of latitudinal shifts of the southern westerly wind belt. Poster U12A-02, F4. *Supplement to EOS Transactions*, AGU volume 80. no. 46 (Holocene wind directions shifted at 4 ka).

Lemke, P., Hibler, W.D., Flato, G., Harder, M. and Kreyscher, M. (1997). On the improvement of sea ice models for climate simulations: the Sea Ice Model Intercomparison Project. *Ann. Glaciol.*; **25** 183–187.

Lund, J.J. and Pedersen, K.R. (1985). Palynology of the marine Jurassic formations in the Vardekloeft ravine, Jameson Land, East Greenland. *Bull. Geol. Soc. Denmark*, **33**: 371–399.

Lutjeharms, J.R.E. (1985). Location of frontal systems between Africa and Antarctica: Some preliminary results. *Deep Sea Res.*, **32**: 329–341.

MacAyeal, D.R. (1993). Binge/purge oscillations of the Laurentide Ice Sheet as a cause of the North Atlantic Heinrich events. *Paleoceanography*, **8**: 775–784.

Mangerud, J. (1991a). The Scandinavian Ice Sheet through the last interglacial/glacial cycle. In: B. Frenzel (Ed.) *Klimageschichtliche Probleme der letzten 130,000 Jahre*, Stuttgart, New York: G. Fischer; 307–330.

Mangerud, J. (1991b). The last interglacial/glacial cycle in northern Europe. In: L. Shane and E. Cushing (Eds) *Quaternary Landscapes*. Minneapolis: Univ. of Minnesota Press; 38–75.

Martin, T. and Angstein, E. (2000). Large-scale drift of Arctic sea ice retrieved from passive microwave satellite data. *J. Geophys. Res.*; **105**: 8775–8788.

Mayewski, P.A., Meeker, L.D., Whitlow, S., Twickler, M.S., Morrison, M.C., Bloomfield, P., Bond, G.C., Alley, R.B., Gow, A.J., Grootes, P.M., Meese, D.A., Ram, M., Taylor, K.C. and Wumkes, W. (1994). Changes in atmospheric circulation and ocean ice cover over the North Atlantic during the last 41,000 years. *Science*, **263**: 1747–1751.

Minicucci, D.A. and Clark, D.L. (1983). A late Cenozoic stratigraphy for glacial-marine sediments of the Eastern Arctic Cordillera, Central Arctic Ocean. In: B.F. Molnia (Ed.) *Glacial-Marine Sedimentation*, New York: Plenum press; 331–365.

Molnia, B.F. (Ed.) (1983). *Glacial-Marine Sedimentation*, New York: Plenum Press.

Mysak, LA. and Venegas, S.A. (1998). Decadal climate oscillations in the Arctic: A new feedback loop for atmosphere-ice-ocean interactions. *Geophysical Research Letters*, **25(19)**: 3607–3610.

Nansen, Fridtjof (1898). *Farthest North: The Norwegian Polar Expedition 1893–1896 Vols 1 and 2* . London: George Newnes Ltd. (Vol. 1 describes sailing through open water and the ice drift passage; Vol. 2 describes the sled trip that almost killed Nansen. Excellent descriptions and measurements of the Arctic in summer and winter, ice thicknesses between 2.3 and 2.8 m; Atlantic layer surface at 220 m water depth, rafted ice floes of great thickness, > 10 m, lots of brown algae, diatoms, etc. in and on the ice, the Arctic is not lifeless at all, plus eye-witness accounts of sediment 'mineral dust' on the ice, plus the menu on board 'Fram' and Ringnes beer!).

Nares, G.S. (1878). *Narrative of a Voyage to the Polar Sea during 1875-6 in H. M. Ships 'Alert' and 'Discovery'*. London: Sampson, Low, Marsten, Searle & Rivington, (description of river sediment onto sea ice).

Nørgaard-Pedersen, N., Spielhagen, R.F., Thiede, J., and Kassens, H. (1998). Central Arctic surface ocean environment during the past 80,000 years. *Paleoceanography*, **13(2)**: 193–204.

Nürnberg, D., Wollenburg, I., Dethleff, D., Eiken, H., Kassens, H., Letzig, T., Reimnitz, E. and Thiede, J. (1994). Sediments in Arctic sea ice – implications for entrainment, transport and release. *Marine Geology*, **119**: 185–214.

Oftedahl, C. (1952). Studies of the igneous rock complex in the Oslo Region. XII. The Lavas. Vid. Skrift., I. mat. naturv. KL., 3, Oslo.

Oftedahl, C. (1980). Geology of Norway. In: *NGU Contributions to the 26th International Geological Congress*. Guide Book, Paris: 7–114.

Paterson, W.S.B. (1994). The *Physics of Glaciers. 3rd edition.* Oxford: Pergamon/ Elsevier Science Ltd.

Peltier, W.R. (1988). Global sea level and earth rotation. *Science*, **240**: 895–901.

Perovich, D.A., Richter-Menge, J.A., Tucker III, W.B., Sturm, M., Grenfell, T.C., Light, B. Maykut, G.A. and Eicken, H. (1999). SHEBA: Snow and ice observations. In: J.M. Grebmeier, L.A. Codispoti and J. Hawkey (organizers), *Arctic System Science (ARCSS). Ocean-Atmosphere-Ice Interactions (OAII)*. Abstract volume of the All Hands Meeting and Planning Workshop, Sponsored by the National Science Foundation, 20–22 October 1999, Virginia Beach: 45 and poster.

Perovich, D.K., Andreas, E.L., Curry, J.A., Eiken, H., Fairall, C.W., Grenfell, T.C., Guest, P.S., Intrieri, J., Kadko, D., Lindsay, R.W., McPhee, M.G., Morison, J., Moritz, R.E., Paulson, C.A., Pegau, W.S. Persson, P.O.G., Pinkel, R., Richter-Menge, J.A., Stanton, T., Stern, H., Sturm, M., Tucker III, W.B. and Uttal, T. (1999). Year on ice gives climate insights. *EOS Transactions*, **80(41)**: 481, 485, 486.

Pfirman, S., Gascard, J.C., Wollenburg, I., Mudie, P. and Abelman, A. (1989). Particle-laden Eurasian Arctic sea ice: Observations from July and August 1987. *Polar Research*; **7**: 59–66.

Pfirman, S., Manfred, L., Wollenburg, I., and Schlosser, P. (1990). Sea ice characteristics and the role of sediment inclusions in deep-sea deposition: Arctic-Antarctic comparisons. In: U. Bleil and J. Thiede (Eds) *Geological History of the Polar oceans: Arctic versus Antarctic*. Dordrecht: Kluwer Academic Publishers; 187–211.

Pfirman, S., Wollenburg, I., Goldschmidt, P., Bischof, J. and Lange, M. (1991). Sediment in Barents Sea Ice. In: G. Krause, J. Meincke and H.J. Schwartz (Eds) *Reports on Polar Research, Scientific Cruise reports of Arctic Expeditions ARK VI/ 1–4 of RV 'Polarstern' in 1989*. Bremerhaven: Alfred Wegener Institute for Polar and Marine Research; 33–35.

Powell, R.D. (1983). Glacial marine sedimentation process and lithofacies of temperate tidewater glaciers, Glacier Bay, Alaska. In: B.F. Molnia (Ed.) *Glacial-Marine Sedimentation*. New York: Plenum Press; 185–232.

Proshutinsky, Andrei Y. and Johnson, Mark A. (1997). The circulation regimes of the wind-driven Arctic Ocean. *J. Geophys. Res.*, **102(C6)**: 12,493–12,514.

Rae, John (1850). *Narrative of an Expedition to the Shores of the Arctic Sea in 1846 and 1847*. London: T. and W. Boone (describes anchor ice formation).

Reimnitz, E. and Barnes, P.W. (1975). Sea ice as a geologic agent on the Beaufort Sea shelf of Alaska. In: J.C. Reed and J.E. Sater (Eds) *The Continental Shelf of the Beaufort Sea*. The Arctic Institute of North America; 301–354.

Reimnitz, E. and Barnes, P.W. (1987a). Sea-ice influence on Arctic coastal retreat. In: N.C. Kraus (Ed.) *Coastal Sediments '87*. Proceedings of a speciality conference on advances in understanding of coastal sediment Processes, New Orleans, LA, Vol. II: 1578–1591.

Reimnitz, E. and Barnes, P.W. (1987). Evidence for sea-ice impact on Icy Reef, Beaufort Sea Coast. In: John P. Galloway and Thomas D. Hamilton (Eds), *Geologic Studies in Alaska by the U.S. Geological Survey during 1987*. U.S. Geological Survey Circular 1016: 190–193.

Reimnitz, E., Barnes, P.W. and Weber, W.S. (1993). Particulate matter in pack ice of the Beaufort Gyre. *Journal of Glaciology*, **39(131)**: 186–198.

Reimnitz, E., Clayton, J., Kempema, E., Payne, J. and Weber, W.S. (1993). Interaction of rising frazil with suspended particles: Tank experiments with applications to nature. *Cold Regions Science and Technology*, **21**: 117–135.

Reimnitz, E., Marincovich, L., McCormick, M. and Briggs, W.M. (1992). Suspension freezing of bottom sediment and biota in the Northwest Passage and implications for Arctic Ocean sedimentation. *Canadian Journal of Earth Sciences*, **29**: 693–703.

Reimnitz, E., McCormick, M., Bischof, J., and Darby, D. (1998). Comparing sea-ice sediment load with Beaufort Sea shelf deposits: Is entrainment selective? *Journal of Sedimentary Research*, **68(5)**: 777–787.

Reimnitz, E., McCormick, M., McDougall, K. and Brouwers, E. (1993). Sediment export by ice rafting from a coastal polynya, Arctic Alaska, U.S.A. *Arctic and Alpine Research*, **25**: 83–98.

Rigor, I. and Ortmeyer, M.(2000). Observations of monthly averaged sea surface pressure, temperature, and ice drift velocities in the Arctic Ocean. Website under construction of the International Arctic Buoy Program, Applied Physics Laboratory, University of Washington. Available for all months beginning in 1979.

Rokoengen, K., Gunsleikrud, T., Lien, R.L., Løflaldli, M., Lise, L., Sindre, E. and Vigran, J.O. (1980). Shallow geology on the continental shelf off Møre and Romsdal: Description of the 1:25,000 Quaternary geological map 6203. *IKU Publ.*, **105**: 1–49.

Ruddiman, W.F. (1977). Late Quaternary deposition of ice-rafted sand in the subpolar North Atlantic (lat. 40° to 65° N). *Geol. Soc. Am. Bull.* v. **88**: 1813–1827.

Sarnthein, M., Stremme, H.E. and Mangini, A. (1986). The Holstein interglaciation: Time stratigraphic position and correlation to stable isotope stratigraphy of deep sea sediments. *Quaternary Research*, **26**: 283–298 (good chronology between marine and European terrestrial, Elsterian ice age equals stages 12 – perhaps 16, etc, ages are close to what I have in the NGS cores, good age calibration scale, old references).

Sarnthein, M. and Altenbach, A.V. (1995). Late Quaternary changes in surface water and deep water masses of the Nordic Seas and north-eastern North Atlantic: A review. *Geologische Rundschau*, **84**: 89–107.

Sarnthein, M., Jansen, E., Arnold, M., Duplessy, J.C., Erlenkeuser, H., Flatoy, A., Hahn, M., Veum, T. and Vogelsang, E. (1992). Delta O-18 time-slice reconstruction of meltwater anomalies at Termination I in the North Atlantic between 50 and 80N. In: E. Bard and W.S. Broecker (Eds) *The Last Deglaciation: Absolute and Radiocarbon Chronologies*. NATO ASI series I, Global Environmental Change, Vol. 2. New York: Springer Verlag Ltd; 183–200.

SAS Institute Inc. (1989). The FASTCLUS procedure. In: *SAS/STAT User's Guide, Version 6, Fourth Edition, Volume 1*. Cary, NC: SAS Institute Inc: 823–850.

Scoresby, William Jr. (1823). *Journal of a Voyage to the Northern Whale-Fishery; Including Researches and Discoveries on the Eastern Coast of West Greenland.* Edinburgh: Archibald Constable and Co.

Siedlecka, A. (1970). Investigations of Permian cherts and associated rocks in southern Spitsbergen. *Norsk Pol. Inst. Skrifter*, **147**.

Sigmond, E.M.O., Gustavson, M. and Roberts, D. (1984). Berggrunnskart over Norge, scale 1:1.000.000, *Norges Geol. Unders.*

Spielhagen, R. (1990). *Die Eisdrift in der Framstraße während der letzten 200.000 Jahre.* Ph. D. thesis, Kiel University.

Spjeldnaes, N. (1981). Ice-rafting, an indication of glaciation? *Science*, **214**: 687–688.

St. John, K.E.K. and Krissek, L.A. (1999). Regional patterns of Pleistocene ice-rafted debris flux in the North Pacific. *Paleoceanography*, **14(5)**: 653–662 (sed. rate and IRD flux rate calculations, IRD in NE Pacific consists of rocks from SE Alaska, sandstones, arkoses, low grade metamorphics, some basic igneous rocks, while the NW Pacific has mafic igneous rocks, chert, and rhodochrosite from eastern Kamtchatka, no IRD in the central North Pacific, sed rates of glacial marine sediment are 2–8 cm/kyr, short history of IRD research in the N Pacific, onset of N Pacific ice rafting = glaciers reaching sea level, at ca. 2.6–2.7 Ma, etc.).

St-Onge, D.(1965). *La géomorphologie de l'île Ellef Ringnes, Territoires du Nord-Ouest,* Canada, Geographical Branch, Paper 38.

Syvitsky, J.P.M. (1989). On the deposition of sediment within glacier-influenced fjords: Oceanographic controls. In: R.D. Powell and A. Elverhøi (Eds) Modern glacimarine environments: Glacial and marine controls of modern lithofacies and biofacies. *Marine Geology* (Special Issue). **85**: 301–329.

Syvitsky, J.P.M. and Hein, F.J. (1991). Sedimentology of an Arctic Basin: Itirbilung Fjord, Baffin Island, Northwest Territories. *Geol. Survey of Canada*, Paper 91–11.

Thiede, J., and Mangerud, J. (1999). New map revises extent of last ice sheet over Barents and Kara Seas. *EOS Transactions*, **80(42)**: 493–494 (last edition like the one I am envisioning, ice mostly over the Barents Sea, Franz Josef Land and Novaya Zemlya, Early to middle Weichselian ice also over much of the Kara Sea. Matches our quartz-dominated IRD from Lomonosov Ridge).

Tozer, E.T. and Thorsteinsson, R. (1964). *Western Queen Elizabeth Islands, Arctic Archipelago*. Geological Survey of Canada, Memoir 332.

Trettin, H.P. (1987). Pearya: A composite terrain with Caledonian affinities in northern Ellesmere Island. Can. J. Earth Sci., **24**: 224–245.

Tucker, W.B. III, Gow, A.J., Meese, D.A., Bosworth, H.W. and Reimnitz, E. (1999). Physical characteristics of summer sea ice across the Arctic Ocean. *Journal of Geophysical Research*, **104(C1)**: 1489–1504.

Voelker, Antje H.L. (1999). *Zur Deutung der Dansgaard/Oeschger Ereignisse in ultra-hochauflösenden Sedimentprofilen aus dem Europäischen Nordmeer*. Berichte-Reports, Institut für Geowissenschaften, Universität Kiel, Nr. 9, 278 S.

Vinje, T., Nordlund, N. and Krambekk, A.A. (1997). Monitoring ice thickness in Fram Strait. *J. Geophys. Res.*; **103**: 10,437–10,450.

Vogelsang, E. (1990). Paläo-Ozeanographie des europäischen Nordmeeres anhand von stabilen C- und O- I-sotopen. PhD. thesis, Kiel University.

Vogt, P.R., Crane, K. and Sundvor, E. (1994). Deep Pleistocene iceberg plowmarks on the Yermak Plateau: Sidescan and 3.5 khz evidence for thick calving ice fronts and a possible marine ice sheet in the Arctic Ocean. *Geology*, **22**: 406–406.

Vogt, P.R., Crane, K. and Sundvor, E. (1995). Deep Pleistocene iceberg plowmarks on the Yermak Plateau: Sidescan and 3.5 khz evidence for thick calving ice fronts and a possible marine ice sheet in the Arctic Ocean. Reply. *Geology*, May 1995: 447–448.

Wadhams, P. (1983). Sea ice thickness distribution in Fram Strait. *Nature*, **305**: 108–111.

Washburn, A.L, (1947). *Reconnaissance Geology of Portions of Victoria Island and Adjacent Regions, Arctic Canada*. Geological Society of America, Memoir 22.

Weinelt, M.S., Sarnthein, M., Vogelsang, E. and Erlenkeuser, H. (1991). Early decay of the Barents Shelf Ice Sheet – spread of stable isotope signals across the eastern Norwegian Sea. *Norsk Geologisk Tidsskrift*, **71**: 137–140.

Westall, F. and Fenner, J. (1990). Polar front fluctuations and the upper Gauss to Brunhes Paleooceanographic record in the southeast Atlantic Ocean. In: U. Bleil and J. Thiede (Eds) *Geological History of the Polar Oceans: Arctic versus Antarctic*. NATO ASI Series C: Mathematical and Physical Sciences – Vol. 308. Dortrecht: Kluwer Academic: 761–782.

Wheeler, P.A., (guest editor). (1997). Topical Studies in Oceanography. 1994 Arctic Ocean Section. *Deep Sea Research, part II, Volume 44, No. 8*: 1483–1757.

Wohlfeil, K. (1983). Verbreitung, Herkunft und Bedeutung der Psephite des Seegebietes zwischen Faeröer und Island. *Meteor Forschungsberichte, Reihe C*, **36**: 31–56.

Woldstedt, P. (1965). *Das Eiszeitalter. Volume 3*. Stuttgart: Enke Verlag (comprehensive treatment of the glaciations of Europe, particularly Germany).

Woldstedt, P. and Duphorn, K. (1974). *Norddeutschland und angrenzende Gebiete im Eiszeitalter*. Stuttgart: Koehler Verlag.

Wolff, Thomas C.W. and Thiede, J. (1991). History of terrigenous sedimentation during the past 10 m.y. in the North Atlantic (ODP Legs 104, 105, and DSDP 81). *Marine Geology*, **101**: 83–102 (sets the intensification of northern hemisphere glaciation to 3.4 Ma, others authors suggest a range from 4.2–2.5 Ma).

Wollenburg, I., and Pfirman, S. (1991). Sediment transport by sea ice in the East Greenland Current. In: G. Krause, J. Meincke and H.J. Schwartz (Eds) *Reports on Polar Research, no. 87. Scientific Cruise reports of Arctic Expeditions ARK VI/1–4 of RV 'Polarstern' in 1989*. Alfred Wegener Institute for Polar and Marine Research, Bremerhaven: 44–47.

Wollenburg, I. 1993. Sedimenttransport durch das arktische Meereis: Die rezente lithogene und biogene Materialfracht, *Reports on Polar Research*, **127**.

Wrangell, F. von (1840). Narrative of an expedition to the Polar Sea, in the years 1820, 1821, 1822, and 1823, edited by E. Sabine. London: James Madden and Co. (describes river ice in the sea).

Index